Taking Sides: Clashing Views
in World Politics, 17/e

by John T. Rourke

ISBN-10: 1259342824   ISBN-13: 9781259342820

Contents

Preface

In the first edition of *Taking Sides: Clashing Views in World Politics*, I wrote of my belief in informed argument: "[A] book that debates vital issues is valuable and necessary. . . . [It is important] to recognize that world politics is usually not a subject of absolute rights and absolute wrongs and of easy policy choices. We all have a responsibility to study the issues thoughtfully, and we should be careful to understand all sides of the debates."

It is gratifying to discover, as indicated by the success of *Taking Sides* over 16 editions, that so many of my colleagues share this belief in the value of a debate-format text.

The format of this edition follows a formula that has proved successful in acquainting students with the global issues that we face and in generating discussion of those issues and the policy choices that address them. This book addresses issues on a wide range of topics in international relations. Each issue has two selections, one pro and one con. Each is also accompanied by an issue introduction, which sets the stage for the debate, provides some background information on each author, and generally puts the issue into its political context. Each issue concludes with an "Exploring the Issue" section that summarizes the debate, gives the reader paths for further investigation, and suggests additional resources that might be helpful. I have also provided relevant Internet site addresses (URLs) following each issue.

I have continued to emphasize issues that are currently being debated in the policy sphere. The authors of the selections are a mix of practitioners, scholars, and noted political commentators. Another point of emphasis over the years has been keeping both the issues and selections current. True to that emphasis, over half the issues in this edition are either new or have new selections.

For the instructors, please note that there are tools available through the publisher for using *Taking Sides* in the classroom. A general guidebook, *Using Taking Sides in the Classroom*, which discusses methods and techniques for integrating the debate-style approach into any classroom setting, is also available. Visit www.mhhe.com/createcentral for all instructor resources.

Taking Sides: Clashing Views in World Politics is only one title in the *Taking Sides* series. If you are interested in seeing the table of contents for any of the other titles, please visit the *Taking Sides* website at www.mhhe.com/createcentral.

Students and other readers should be aware that the debates in this book are not one-sided. Each author strongly believes in his or her position. And if you read the debates without prejudging them, you will see that each author makes cogent points. An author may not be "right," but the arguments made in an essay should not be dismissed out of hand, and you should work to remain tolerant of those who hold beliefs that are different from your own. There is an additional consideration to keep in mind as you pursue this debate approach to world politics. To consider divergent views objectively does not mean that you have to remain forever neutral. In fact, once you are informed, you ought to form convictions. More important, you should try to influence international policy to conform better with your beliefs. Write letters to policy-makers; donate to causes you support; work for candidates who agree with your views; join an activist organization. Do something, whichever side of an issue you are on!

John T. Rourke
University of Connecticut, Storrs

Editor of This Volume

JOHN T. ROURKE is professor emeritus and head of the Department of Political Science at the University of Connecticut, Storrs. In addition to this seventeenth edition and the 16 previous editions of *Takings Sides: Clashing Issues in World Politics*, he is the author/editor of *International Politics on the World Stage* (McGraw-Hill, 12 editions), *You Decide! Current Debates in American Politics* (Pearson, 9 editions), *Taking Sides: Clashing Views on Controversial Issues in American Foreign Policy* (McGraw-Hill, 3 editions), *America's Politics: A Diverse Country in a Globalizing World* (Paradigm), *Making Foreign Policy: United States, Soviet Union, China* (Harcourt), *Congress and the Presidency in U.S. Foreign Policymaking: A Study of Interaction and Influence, 1945–1982* (Westview), and *Presidential Wars and American Democracy: Rally 'Round the Chief* (Paragon), and the co-author of (with and Ralph G. Carter and Mark A. Boyer) *Making American Foreign Policy* (Dushkin, 2 editions), and (with Richard P. Hiskes and Cyrus Ernesto Zirakzadeh), *Direct Democracy and International Politics: Deciding International Issues Through Referendums* (Lynne Reinner). He has also published numerous articles in scholarly journals and chapters in edited books.

Academic Advisory Board Members

Members of the Academic Advisory Board are instrumental in the final selection of articles for *Taking Sides* books and ExpressBooks. Their review of the articles for content, level, and appropriateness provides critical direction to the editors and staff. We think that you will find their careful consideration reflected in this book.

Introduction

Some years ago, the Rolling Stones recorded "Sympathy for the Devil." If you have never heard it, it is worth listening to. You can find free video recordings on YouTube, Dailymotion, and other sites. The theme of the song is echoed in a wonderful essay by Marshall Berman, "Have Sympathy for the Devil" (*New American Review*, 1973). The common theme of the Stones' and Berman's works is based on Johann Goethe's Faust. In that classic drama, the protagonist, Dr. Faust, trades his soul to gain great power. He attempts to do good, but in the end he commits evil by, in contemporary paraphrase, "doing the wrong things for the right reasons." Does that make Faust evil, the personification of the devil Mephistopheles among us? Or is the good doctor merely misguided in his effort to make the world better as he saw it and imagined it might be? The point that the Stones and Berman make is that it is important to avoid falling prey to the trap of many zealots who are so convinced of the truth of their own views that they feel righteously at liberty to condemn those who disagree with them as stupid or even diabolical.

It is to the principle of rational discourse, of tolerant debate, that this reader is dedicated. There are many issues in this volume that appropriately excite passion—for example, Issue 7 on whether or not Israel should agree to an independent Palestinian state.

As you will see, each of the authors in all the debates strongly believes in his or her position. If you read these debates objectively, you will find that each side makes cogent points. They may or may not be right, but they should not be dismissed out of hand. It is important to repeat that the debate format does not imply that you should remain forever neutral. In fact, once you are informed, you ought to form convictions, and you should try to act on those convictions and try to influence international policy to conform better with your beliefs. Ponder the similarities in the views of two very different leaders, a very young president in a relatively young democracy and a very old emperor in a very old country: In 1963 President John F. Kennedy, in recalling the words of the author of the epic poem *The Divine Comedy* (1321), told a West German audience, "Dante once said that the hottest places in hell are reserved for those who in a period of moral crisis maintain their neutrality." That very same year, while speaking to the United Nations, Ethiopia's emperor Haile Selassie (1892–1975) said, "Throughout history it has been

the inaction of those who could have acted, the indifference of those who should have known better, the silence of the voice of justice when it mattered most that made it possible for evil to triumph."

The point is: Become informed. Then do something! Write letters to policymakers, donate money to causes you support, work for candidates with whom you agree, join an activist organization, or any of the many other things that you can do to make a difference. What you do is less important than that you do it.

Approaches to Studying International Politics

As will become evident as you read this volume, there are many approaches to the study of international politics. Some political scientists and most practitioners specialize in substantive topics, and this reader is organized along topical lines.

Many of the debates in *Taking Sides: Clashing Views in World Politics* focus on regional and country-specific issues, including the U.S. positions in the world, trends in Russian domestic and foreign policy, and the growing power and assertiveness of China. Then there are issues such as who is responsible for the lack of peace between Israelis and Palestinians, and the degree to which the policies of President Obama have eased or worsened the ongoing turmoil of the Middle East and the threat of terrorism, the external manifestation of that turmoil.

The economic strength of individual countries and the flow of resources, funds, and other forms of internal economic exchange also play a powerful role in global affairs. Can, for example, the United States afford to continue to play the preeminent role in world politics? Also in question is the future of the European Union, the bold experiment by that body to create a unified economy and at least a degree of political unity that has been underway for sixty years. There are also growing questions about the benefits and drawbacks of an even more sweeping change in international economics, globalization.

An unfortunate continuing aspect of world affairs is violence through war and terrorism. It is a testament to how much these sad phenomena are present on the world stage that a third or more of the debates in this volume touch on preparing for, using, or preventing violence. The scope of these issues ranges from weapons of mass destruction and whether, for instance, the world should agree to stop testing nuclear weapons, to such more precise applications of violence, such as using drones to target and kill enemy leaders.

Yet another key aspect of world politics is the effort to create order in the international system by promoting international organizations, such as the United Nations, and by promoting international law. A central actor in this effort is the United Nations, but its ability to meet its goals is severely constrained, leaving supporters arguing the UN would be further empowered and critics dismissing it as fundamentally flawed. International law is much less widely and powerfully enforced than domestic law within countries. There are many people, though, who argue international law should be expanded to cover such topics as the human rights of women and other historically marginalized and ill treated groups. Adding "bite" to international law enforcement through such methods as the recent creating of the International Criminal Court is also a controversial step.

People around the world have also come to understand that the Earth is both finite in its resources and subject to attack by pollution, chemical degradation of the ozone layer, and the increasing levels of carbon dioxide and other greenhouse gases in the atmosphere. Over the past few decades these threats have been met in two related ways. One has been an effort to reach an international agreement to reduce or reverse global warming. There has been little progress on this front. The second approach is for individual countries, especially those that emit the most pollution, greenhouse gases, and others and those that consume the most resources overall or per capita, implement domestic changes to increase energy sources that emit no or fewer greenhouse gases, conserve on resource consumption, or otherwise achieve a level of activity that allows the Earth's environment and resources to be sustained.

Dynamics of World Politics

The action on the global stage today is vastly different from what it was a few decades ago, or even a few years ago. Technology is one of the causes of this change. Technology has changed communications, manufacturing, health care, and many other aspects of the human condition. Technology has given humans the ability to create biological, chemical, and nuclear compounds and other materials that in relatively small amounts have the ability to kill and injure huge numbers of people. Another negative by-product of technology may be the vastly increased consumption of petroleum and other natural resources and the global environmental degradation that has been caused by discharges of waste products, deforestation, and a host of other technology-enhanced human activities. Technology has also changed warfare in many ways. Some of these, like the use of drones to attack enemies away from the immediate battlefield, raise new matters of law and morality that need to be addressed. Technology is also a key element of Issue 11, which deals with the ability to ensure the stability of the current U.S. nuclear arms arsenal and the ability to verify that other countries are not secretly testing nuclear weapons as well as whether the United States should agree to never again test nuclear weapons.

Technological changes also highlight the increased role of economics in world politics. Economics has always played a role, but traditionally the main focus has been on strategic-political questions—especially military power. This concern still strongly exists, but now it shares the international spotlight with economic issues. One important change in recent decades has been the rapid growth of regional and global markets and the promotion of free trade and other forms of international economic interchange. As noted earlier, the world's most advanced regional effort at governance, the European Union, is experiencing some difficulties. These may bode ill for the EU's future or may be just "growing pains" or the ups and downs of any country or international organization. Much is said about globalization and whether it has basic flaws or if reform can address current issues.

Another change in the world system has to do with the main international actors. At one time, states (countries) were practically the only international actors on the world stage. Now, and increasingly so, there are other actors. The EU and some other actors are regional. Others, such as the United Nations, are global actors. International law and its courts, like the ICC, are also global actors. Now, for the first time, individuals are being held accountable for war crimes and other violations of international law by neutral international courts, not just by courts created by victorious countries to punish the leaders of defeated countries.

Perceptions Versus Reality

In addition to addressing the general changes in the world system outlined above, the debates in this reader explore the controversies that exist over many of the fundamental issues that face the world. One key to these debates is the differing perceptions that protagonists bring to them. There may be a reality in world politics, but very often that reality is obscured. Many observers, for example, are alarmed by the seeming rise in radical actions by Islamic fundamentalists. However, the image of Islamic radicalism is not a fact but a perception, perhaps correct, perhaps not. In cases such as this, though, it is often the perception, not the reality that is more important because policy is formulated on what decision makers think, not necessarily on what is. Thus, perception becomes the operating guide, or operational reality, whether it is true or not. Perceptions result from many factors. One factor is the information that decision makers receive. For a variety of reasons, the facts and analyses that are given to leaders are often inaccurate or represent only part of the picture. The conflicting perceptions of Israelis and Palestinians, for example, make the achievement of peace in Israel very difficult. Many Israelis and Palestinians fervently believe that the conflict that has occurred in the region over the past 50 years is the responsibility of the other. Both sides also believe in the righteousness of their own policies. Even if both sides are well-meaning, the perceptions of hostility that each holds mean that the operational reality often has to be violence. These differing perceptions are a key element in the debate in Issue 3 about whether China is becoming a dangerous superpower.

A related aspect of perception is the tendency to see oneself differently than some others do. Specifically, the tendency is to see oneself as benevolent and to perceive rivals as sinister. Americans, Western Europeans, and other increasingly see Russia under President Putin as a growing menace. It is difficult to look at events through the eyes of the Russians and see how the expansion of the European Union and North Atlantic Treaty Organization to countries on or near Russia's borders might seem threatening to Russians. Understanding the Russian viewpoint does not necessarily mean agreeing with Russia's action, but it will help in shaping a wise response. As Chinese strategist Sun Tzu advised sagely some 2,700 years ago in *The Art of War*, "know your enemy."

Levels of Analysis

Political scientists approach the study of international politics from three levels of analysis: system, state, and individual. The most macroscopic view is system-level analysis. This is a top-down approach that maintains that world factors virtually compel countries to follow certain foreign policies. Governing factors include the number of powerful actors, geographic relationships, economic needs, and technology. System analysts hold that a country's internal political system and its leaders do not have a major impact on policy. As such, political scientists who work from this perspective are interested in exploring the governing factors, how they cause policy, and how and why systems change.

One key factor in how a system works is the number of major powers (or poles) in it. Prior to World War II, the world had a multipolar (four or more major powers) system, with Great Britain, Germany, Japan, the Soviet Union, and the United States as its poles. After the end of World War II, the world was structured as a bipolar system, dominated by the United States and the Soviet Union. Furthermore, each superpower was supported by a tightly organized and dependent group of allies. For a variety of reasons, including changing economics and the nuclear standoff, the bipolar system has faded. Some political scientists argue that the bipolar system is being replaced by a multipolar system. In such a configuration, those who favor balance-of-power politics maintain that it is unwise to ignore power considerations. Is it possible that the future holds a tripolar system? What the polar future brings will be based in part on the willingness and capability of the United States to play the role of the "indispensable country," the most powerful and globally engaged country. The policies and capabilities of Russia and China will also play a key role in shaping the future's polar structure. It may also be that other powers, a politically united Europe, an emerging India, a revitalized and rearmed Japan, or some rising power(s) could create a multipolar system, akin to what existed prior to World War II.

System-level analysis also deals with global actors and issues. World forces are evident in such issues as the viability of the UN, how far and on what topics the role of international law should increase. The nuclear age has also affected the operation of the international system, as has technology, such as rapid global communications, global ideas such as human rights, and worldwide phenomena such as global warming.

State-level analysis is the middle and most common level of analysis. Social scientists who study world politics from this perspective focus on how countries, singly or comparatively, make foreign policy. In other words, this perspective is concerned with internal political dynamics, such as the roles of and interactions between the executive and legislative branches of government, the impact of bureaucracy, the role of interest groups, and the effect of public opinion. This level of analysis is very much in evidence in such issues as the political barriers within the United States, and particularly in the U.S. Senate, to U.S ratification of the Comprehensive (Nuclear) Test Ban Treat, the Convention to Eliminate All Forms of Discrimination Against Women, and the Treaty of Rome creating the ICC. Also an illustration of state-level analysis is the struggle in the Untied States over whether to adopt unilateral restrictions on the emission of greenhouse gases between a determined President Obama and a reluctant Congress.

People as individuals also help shape world politics. The focus on individual-level analysis ranges from top (leaders), such as the mindset of Russia's President Vladimir Putin, to the bottom (individual citizens), such as the decision of Edward Snowden to use his position as an American intelligence analyst to gather and make public a huge catalog of classified communications and other documents.

Realism and Idealism

Another way for students and practitioners of world politics to approach their subject is to focus on what is called the realist versus the idealist (or liberal) debate. Realists tend to assume that the world is permanently flawed and therefore advocate following policies in their country's narrow self-interests. Idealists take the approach that the world condition can be improved substantially by following policies that, at least in the short term, call for some risk or self-sacrifice. This divergence is an element of many of the debates in this book. Right now, for example, the United States refuses to recognize the authority of the International Criminal Court over any U.S. citizen. Certainly, allowing Americans who commit war crimes to be tried there would diminish U.S. sovereign independence, but some people argue that the ICC is a major step toward a more just world and does not pose any threat to the United States.

The Political and Ecological Future

Future world alternatives are discussed in many of the issues in this volume. Abraham Lincoln once said, "A house divided against itself cannot stand." One suspects that the sixteenth president might say something similar about the world today if he were with us. Issue 1, for example, debates whether growing globalization is a positive or negative trend. The world has responded to globalization by creating and strengthening the UN, the International Monetary Fund, the World Bank, the World Trade Organization, and many other international organizations to try to regulate the increasing number of international interactions. There can be little doubt that the role of global governance is growing, and this reality is the spark behind specific debates about the future that are taken up in many of the

selections. Far-reaching alternatives to a state-centric system based on sovereign countries include international organizations, whether global like the UN or regional like the EU, taking over some (or all) of the sovereign responsibilities of national governments. The global future also involves the ability of the world to prosper economically while not denuding itself of its natural resources or destroying the environment.

Increased Role of Economics

Economics has always played a part in international relations, but the traditional focus has been on strategic political affairs, especially questions of military power. Now, however, political scientists are increasingly focusing on the international political economy, or the economic dimensions of world politics. International trade, for instance, has increased dramatically, expanding from an annual world export total of $20 billion in 1933 to $15 trillion in 2011. The impact has been profound. The domestic economic health of most countries is heavily affected by trade and other aspects of international economics. Since World War II there has been an emphasis on expanding free trade by decreasing tariffs and other barriers to international commerce. In recent years, however, a downturn in the economies of many of the industrialized countries has increased calls for more protectionism. Yet restrictions on trade and other economic activity can also be used as diplomatic weapons. The intertwining of economies and the creation of organizations to regulate them, such as the World Trade Organization, is raising issues of sovereignty and other concerns.

Conclusion

Having discussed many of the various dimensions and approaches to the study of world politics, it is incumbent on this editor to advise against your becoming too structured by them. Issues of focus and methodology are important both to studying international relations and to understanding how others are analyzing global conduct. However, they are also partially pedagogical. In the final analysis, world politics is a highly interrelated, perhaps seamless, subject. No one level of analysis, for instance, can fully explain the events on the world stage. Instead, using each of the levels to analyze events and trends will bring the greatest understanding.

Similarly, the realist-idealist division is less precise in practice than it may appear. As some of the debates indicate, each side often stresses its own standards of morality.

Which is more moral: defeating a dictatorship or sparing the sword and saving lives that would almost inevitably be lost in the dictator's overthrow? Furthermore, realists usually do not reject moral considerations. Rather, they contend that morality is but one of the factors that a country's decision makers must consider. Realists are also apt to argue that standards of morality differ when dealing with a country as opposed to an individual. By the same token, most idealists do not completely ignore the often dangerous nature of the world. Nor do they argue that a country must totally sacrifice its short-term interests to promote the betterment of the current and future world. Thus, realism and idealism can be seen most accurately as the ends of a continuum—with most political scientists and practitioners falling somewhere between, rather than at, the extremes. The best advice, then, is this: think broadly about international politics. The subject is very complex, and the more creative and expansive you are in selecting your foci and methodologies, the more insight you will gain. To end where we began, with Dr. Faust, I offer his last words in Goethe's drama, *"Mehr licht"* . . . More light! That is the goal of this book.

John T. Rourke
University of Connecticut, Storrs

Unit 1

UNIT

Major Power Issues

*N*ot all countries are equal, at least not in their impact on the international system. As noted in the introduction, a key determinant of how international system works is the number of major powers (or poles). Today the polar system is tripolar. Being a pole does not imply parity among the poles. Of the current three poles, the United States is the most powerful. China is on the rise, however. So is Russia, after a short eclipse from even polar status after the collapse of the Soviet Union. A central operation in a tripolar system is for each of the three powers to work to try at minimum to ensure that it is not the "odd man out," that is, alone versus in an alliance of the other two poles. The optimal position is for a country to have reasonable relations with both other poles while they are at odds with one another.

In addition to the direct dynamics among the poles, the fact that each is powerful means that the specific policies of each powerful country have a greater impact on the international system than other countries.

Also note that polar systems are not stable. Power ebbs and flows, and countries sometimes gain or lose status as a major power. India and Japan may well be on the road to becoming major powers, and some see Brazil as a possible major power in the more distant future.

Selected, Edited, and with Issue Framing Material by:
John T. Rourke, *University of Connecticut, Storrs*

ISSUE

Should the United States Seek to Remain the "Indispensable" Country?

YES: Hillary Clinton, from "Remarks on American Leaderships," Remarks Delivered at the Council on Foreign Relations, Washington, D.C. (2013)

NO: Ted Galen Carpenter, from "Delusions of Indispensability," *The National Interest* (2013)

Learning Outcomes

After reading this issue, you will be able to:

- Describe what "indispensable country" means.
- Explain why Secretary of State Clinton and others want the United States to continue to be the indispensable country.
- Explain the drawbacks to trying to be the indispensable country.
- Lay out the range of options the United States has between the extremes of trying to be a unipolar power and isolationism.

ISSUE SUMMARY

YES: Hillary Clinton, during remarks to the Council on Foreign Relations just before leaving her post as U.S. Secretary of State, tells her audience that the United States remains the world's most important power and that it is important that the United States continue to play the role of the globe's indispensable state. Without too specifically delving into political science theory, she takes the view that an international system (state of affairs) without a central power to order and police those affairs to at least some degree is likely to become chaotic and conflictive.

NO: Ted Galen Carpenter, a respected and well-published scholar who is a senior fellow at the Cato Institute, a libertarian think-tank, argues that the United States does not have the resources or the national will to try to play the indispensable country. He maintains that those like Clinton who advocate that role put the United States at risk of overextending itself and suffering financial costs, policy failures, and other unpleasant outcomes.

T he emergence of the United States as a world power began in the late 1800s. Rapid industrialization was one factor. By 1900, for example, U.S. steel production was equal to that of the British and French combined. A rapidly growing population, largely due to high immigration, also added to U.S. power, with the U.S. population in 1890 equal to the combined populations of Great Britain and France. A third factor, the willingness of Americans to be involved in world affairs, also began to change. One indicator of declining isolation is U.S. involvement in the Spanish–American War (1898), World War I beginning in 1917, World War II beginning in 1941, and a range of minor conflicts.

World War II devastated most of the existing major powers. However, U.S. industrial capacity and infrastructure survived untouched, and in 1946 the U.S. gross national product (GNP) was a stunning 50 percent of the GNP of the entire world. Militarily, U.S. air and naval power were the world's best, and the United States was the only country with atomic weapons. Furthering U.S. power even more, Americans were willing to be active in world affairs. Symbolizing this, the United States joined the UN (1945), whereas the country had refused to be part of the League of Nations after World War I. In 1946, the United States was not only the world's most powerful country, but in its own class, that of superpower.

This sole-superpower status soon changed with the emergence of the Soviet Union as a rival superpower. While the United States had rapidly demobilized most of its army after 1945, the Soviets had maintained an immense ground force, one that could threaten Western Europe. The Soviets also had a hostile ideology, communism, and became even more potent in 1949 when they acquired atomic weapons.

What ensued for the next several decades was the cold war, with two superpowers in a bipolar confrontation. Then, for reasons that are still debated, the Soviet Union collapsed in 1991. Just a year before amid the decay of Soviet power, analyst Charles Krauthammer had heralded the coming of a "unipolar moment" with the United States at "the center of world power" as "the unchallenged superpower." He was right, and in 1992 the United States once again, as had occurred in the late 1940s, stood alone as the world's only superpower.

Still, the moment was different than it had been in 1946. Europe had long ago recovered economically, and many of its countries had joined an economic powerhouse, the European Union. Japan had also become an economic power, and China was fast approaching that status. The United States was still by far the world's largest economy, but its share of the world GDP had declined from about half in 1946 to 23 percent in 1991. The U.S. share of the world economy has remained basically stable since then, and stood at 22 percent in 2013. Still, there are numerous signs of trouble in the U.S. economy. It is just recovering from a severe recession that began in 2008, and the chronic and immense federal budget deficit in 2014 will be about $650 billion, although that is down by about half from what it was only recently. One big change has been the decline in the costs of conducting the wars in Iraq and Afghanistan.

U.S. military power remains beyond that of any other country, but it was severely stretched by the difficulty of maintaining U.S. global commitment while fighting in Iraq and Afghanistan. Those two wars also showed how difficult it is to apply conventional military strength to irregular warfare, much less terrorism. It is also the case that budget cuts to help trim deficit spending will necessarily diminish U.S. capabilities. Finally, the seeming endless conflicts in Iraq and Afghanistan have left Americans less certain they are willing to pay the price of being the indispensible country. Secretary of State Hillary Clinton argues that the benefits are worth the burdens of continuing to play that role. Cato Institute scholar Ted Galen Carpenter disagrees, taking the position that the United States will be better off if it understands its limitations and does not risk negative domestic and international consequences by trying to be the world central power.

YES ⬅

Hillary Rodham Clinton

Remarks on American Leaderships

I want to thank the board of the Council on Foreign Relations and all my friends and colleagues and other interested citizens who are here today, because you respect the Council, you understand the important work that it does, and you are committed to ensuring that we chart a path to the future that is in the best interests not only of the United States, but of the world.

Tomorrow is my last day as Secretary of State. And though it is hard to predict what any day in this job will bring, I know that tomorrow, my heart will be very full. Serving with the men and women of the State Department and USAID has been a singular honor. And [the new] Secretary [of State John] Kerry will find there is no more extraordinary group of people working anywhere in the world. So these last days have been bittersweet for me, but this opportunity that I have here before you gives me some time to reflect on the distance that we've traveled, and to take stock of what we've done and what is left to do.

I think it's important [to note] what we faced in January of 2009: Two wars, an economy in freefall, traditional alliances fraying, our diplomatic standing damaged, and around the world, people questioning America's commitment to core values and our ability to maintain our global leadership. That was my inbox on day one as your Secretary of State.

Today, the world remains a dangerous and complicated place, and of course, we still face many difficult challenges. But a lot has changed in the last four years. Under President [Barack] Obama's leadership, we've ended the war in Iraq, begun a transition in Afghanistan, and brought Usama bin Ladin to justice. We have also revitalized American diplomacy and strengthened our alliances. And while our economic recovery is not yet complete, we are heading in the right direction. In short, America today is stronger at home and more respected in the world. And our global leadership is on firmer footing than many predicted.

To understand what we have been trying to do these last four years, it's helpful to start with some history. Last year, I was honored to deliver the Forrestal Lecture at the Naval Academy, named for our first Secretary of Defense after World War II. In 1946, James Forrestal noted in his diary that the Soviets believed that the post-war world should be shaped by a handful of major powers acting alone. But, he went on, "The American point of view is that all nations professing a desire for peace and democracy should participate."

And what ended up happening in the years since is something in between. The United States and our allies succeeded in constructing a broad international architecture of institutions and alliances—chiefly the UN [United Nations], the IMF [International Monetary Fund], the World Bank, and NATO [North Atlantic Treaty Organization]—that protected our interests, defended universal values, and benefitted peoples and nations around the world. Yet it is undeniable that a handful of major powers did end up controlling those institutions, setting norms, and shaping international affairs.

Now, two decades after the end of the Cold War, we face a different world. More countries than ever have a voice in global debates. We see more paths to power opening up as nations gain influence through the strength of their economies rather than their militaries. And political and technological changes are empowering non-state actors, like activists, corporations, and terrorist networks.

At the same time, we face challenges, from financial contagion to climate change to human and wildlife trafficking, that spill across borders and defy unilateral solutions. As President Obama has said, the old postwar architecture is crumbling under the weight of new threats. So the geometry of global power has become more distributed and diffuse as the challenges we face have become more complex and crosscutting.

So the question we ask ourselves every day is: What does this mean for America? And then we go on to say: How can we advance our own interests and also uphold a just, rules-based international order, a system that does provide clear rules of the road for everything from intellectual property rights to freedom of navigation to fair labor standards?

Hillary Rodham Clinton. Remarks Delivered at the Council on Foreign Relations, January 31, 2013.

Simply put, we have to be smart about how we use our power. Not because we have less of it—indeed, the might of our military, the size of our economy, the influence of our diplomacy, and the creative energy of our people remain unrivaled. No, it's because as the world has changed, so too have the levers of power that can most effectively shape international affairs.

I've come to think of it like this: [President Harry S] Truman and [Secretary of State Dean G.] Acheson were building the Parthenon with classical geometry and clear lines [in the late 1940s]. The pillars were a handful of big institutions and alliances dominated by major powers. And that structure delivered unprecedented peace and prosperity. But time takes its toll, even on the greatest edifice.

And we do need a new architecture for this new world. Where once a few strong columns could hold up the weight of the world, today we need a dynamic mix of materials and structures.

Now, of course, American military and economic strength will remain the foundation of our global leadership. As we saw from the intervention to stop a massacre in Libya to the raid that brought bin Ladin to justice, there will always be times when it is necessary and just to use force. America's ability to project power all over the globe remains essential.

And I'm very proud of the partnerships that the State Department has formed with the Pentagon [and its leaders]. By the same token, America's traditional allies and friends in Europe and East Asia remain invaluable partners on nearly everything we do. And we have spent considerable energy strengthening those bonds over the past four years.

And, I would be quick to add, the UN, the IMF, the World Bank, and NATO are also still essential. But all of our institutions and our relationships need to be modernized and complemented by new institutions, relationships, and partnerships that are tailored for new challenges and modeled to the needs of a variable landscape, like how we elevated the G-20 [the Group of 20: the world 20 of the leading economic powers] during the financial crisis, or created the Climate and Clean Air Coalition out of the State Department to fight short-lived pollutants like black carbon, or worked with partners like Turkey, where the two of us stood up the first Global Counterterrorism Forum.

We're also working more than ever with invigorated regional organizations. Consider the African Union in Somalia and the Arab League in Libya, even sub-regional groups like the Lower Mekong Initiative that we created to help reintegrate Burma into its neighborhood and try to work across national boundaries on issues like whether dams should or should not be built.

We're also, of course, thinking about old-fashioned shoe-leather diplomacy in a new way. I have found it, and I've said this before, highly ironic that in today's world, when we can be anywhere virtually, more than ever, people want us to actually show up. But while a Secretary of State in an earlier era might have been able to focus on a small number of influential capitals, shuttling between the major powers, today we, by necessity, must take a broader view.

And people say to me all the time, "I look at your travel schedule; why Togo?" Well, no Secretary of State had ever been to Togo. But Togo happens to hold a rotating seat on the UN Security Council. Going there, making the personal investment has a strategic purpose.

And it's not just where we engage, but with whom. You can't build a set of durable partnerships in the 21st century with governments alone. The opinions of people now matter as to how their governments work with us, whether it's democratic or authoritarian. So in virtually every country I have visited, I've held town halls and reached out directly to citizens, civil society organizations, women's groups, business communities, and so many others. They have valuable insights and contributions to make. And increasingly, they are driving economic and political change, especially in democracies.

The State Department now has Twitter feeds in 11 languages. And just this Tuesday, I participated in a global town hall and took questions from people on every continent, including, for the first time, Antarctica.

So the point is: We have to be strategic about all the levers of global power and look for the new levers that could not have been possible or had not even been invented a decade ago. We need to widen the aperture of our engagement, and let me offer a few examples of how we're doing this.

First, technology. You can't be a 21st century leader without 21st century tools, not when people organize pro-democracy protests with Twitter and while terrorists spread their hateful ideology online. That's why I have championed what we call 21st century statecraft.

We've launched an interagency Center for Strategic Counterterrorism Communications at State. Experts, tech-savvy specialists from across our government fluent in Urdu, Arabic, Punjabi, Somali, use social media to expose al-Qaida's contradictions and abuses, including its continuing brutal attacks on Muslim civilians.

We're leading the effort also to defend internet freedom so it remains a free, open, and reliable platform for everyone. We're helping human rights activists in oppressive internet environments get online and communicate more safely. Because the country that built the internet ought to be leading the fight to protect it from those who would censor it or use it as a tool of control.

Second, our nonproliferation agenda. Negotiating the New START [Strategic Arms Reduction Talks] Treaty with Russia was an example of traditional diplomacy at its best. Then working it through the Congress was an example of traditional bipartisan support at its best. But we also have been working with partners around the world to create a new institution, the Nuclear Security Summit, to keep dangerous materials out of the hands of terrorists. We conducted intensive diplomacy with major powers to impose crippling sanctions against Iran and North Korea. But to enforce those sanctions, we also enlisted banks, insurance companies, and high-tech international financial institutions. And today, Iran's oil tankers sit idle, and its currency has taken a massive hit.

Now, this brings me to a third lever: economics. Everyone knows how important that is. But not long ago, it was thought that business drove markets and governments drove geopolitics. Well, those two, if they ever were separate, have certainly converged. So creating jobs at home is now part of the portfolio of diplomats abroad. They are arguing for common economic rules of the road, especially in Asia, so we can make trade a race to the top, not a scramble to the bottom. We are prioritizing economics in our engagement in every region, like in Latin America, where, as you know, we ratified free trade agreements with Colombia and Panama.

And we're also using economic tools to address strategic challenges, for example, in Afghanistan, because along with the security transition and the political transition, we are supporting an economic transition that boosts the private sector and increases regional economic integration. It's a vision of transit and trade connections we call the New Silk Road.

A related lever of power is development. And we are helping developing countries grow their economies not just through traditional assistance, but also through greater trade and investment, partnerships with the private sector, better governance, and more participation from women. We think this is an investment in our own economic future. And I love saying this, because people are always quite surprised to hear it: Seven of the 10 fastest growing economies in the world are in Africa. Other countries are doing everything they can to help their companies win contracts and invest in emerging markets. Other countries still are engaged in a very clear and relentless economic diplomacy. We should too, and increasingly, we are.

And make no mistake: There is a crucial strategic dimension to this development work as well. Weak states represent some of our most significant threats. We have an interest in strengthening them and building more capable partners that can tackle their own security problems at home and in their neighborhoods, and economics will always play a role in that.

Next, think about energy and climate change. Managing the world's energy supplies in a way that minimizes conflict and supports economic growth while protecting the future of our planet is one of the greatest challenges of our time. So we're using both high-level international diplomacy and grassroots partnerships to curb carbon emissions and other causes of climate change. We've created a new bureau at the State Department focused on energy diplomacy as well as new partnerships like the U.S.-EU Energy Council. We've worked intensively with the Iraqis to support their energy sector, because it is critical not only to their economy, but their stability as well. And we've significantly intensified our efforts to resolve energy disputes from the South China Sea to the eastern Mediterranean to keep the world's energy markets stable. Now this has been helped quite significantly by the increase in our own domestic production. It's no accident that as Iranian oil has gone offline because of our sanctions, other sources have come online, so Iran cannot benefit from increased prices.

Then there's human rights and our support for democracy and the rule of law, levers of power and values we cannot afford to ignore. In the last century, the United States led the world in recognizing that universal rights exist and that governments are obligated to protect them. Now we have placed ourselves at the frontlines of today's emerging battles, like the fight to defend the human rights of the LGBT [lesbian, gay, bisexual, transgender] communities around the world and religious minorities wherever and whoever they are. But it's not a coincidence that virtually every country that threatens regional and global peace is a place where human rights are in peril or the rule of law is weak.

More specifically, places where women and girls are treated as second-class, marginal human beings. Just ask young Malala Yousafzai from Pakistan. Ask the women of northern Mali who live in fear and can no longer go to school. Ask the women of the Eastern Congo who endure rape as a weapon of war. [Malala Yousafzai is an activist for women's rights who from age 12 in 2009 wrote a blog detailing the Taliban's oppression in Pakistan, including suppression of girls' right to go to school. Her writing gained international note. In 2012, gunmen boarded Malala's school bus and shot her in the head, nearly killing her. Global outrage followed. She has substantially recovered from her wounds and has been honored widely.

She spoke at the UN on her 16th birthday, July 12, 2013, and urged for worldwide access to education.]

And that is the final lever that I want to highlight briefly. Because the jury is in, the evidence is absolutely indisputable: If women and girls everywhere were treated as equal to men in rights, dignity, and opportunity, we would see political and economic progress everywhere. So this is not only a moral issue, which, of course, it is. It is an economic issue and a security issue, and it is the unfinished business of the 21st century. It therefore must be central to U.S. foreign policy.

One of the first things I did as Secretary was to elevate the Office of Global Women's Issues under the first Ambassador-at-Large, Melanne Verveer. And I'm very pleased that yesterday, the President signed a memorandum making that office permanent. In the past four years, we've made a major push at the United Nations to integrate women in peace and security-building worldwide, and we've seen successes in places like Liberia. We've urged leaders in Egypt, Tunisia, and Libya to recognize women as equal citizens with important contributions to make. We are supporting women entrepreneurs around the world who are creating jobs and driving growth.

So technology, development, human rights, women. Now, I know that a lot of pundits hear that list and they say: Isn't that all a bit soft? What about the hard stuff? Well, that is a false choice. We need both, and no one should think otherwise.

I will be the first to stand up and proclaim loudly and clearly that America's military might is and must remain the greatest fighting force in the history of the world. I will also make very clear, as I have done over the last years, that our diplomatic power, the ability to convene, our moral suasion is effective because the United States can back up our words with action. We will ensure freedom of navigation in all the world's seas. We will relentlessly go after al-Qaida, its affiliates, and its wannabes. We will do what is necessary to prevent Iran from obtaining a nuclear weapon.

There are limits to what soft power [such as diplomacy, information, and cultural relations] on its own can achieve. And there are limits to what hard power on its own can achieve. That's why, from day one, I've been talking about smart power. And when you look at our approach to two regions undergoing sweeping shifts, you can see how this works in practice.

First, America's expanding engagement in the Asia Pacific. Now, much attention has been focused on our military moves in the region. And certainly, adapting our force posture is a key element of our comprehensive strategy. But so is strengthening our alliances through new economic

and security arrangements. We've sent Marines to Darwin, but we've also ratified the Korea Free Trade Agreement. We responded to the triple disaster in Japan through our governments, through our businesses, through our not-for-profits, and reminded the entire region of the irreplaceable role America plays.

First and foremost, this so-called pivot has been about creative diplomacy:

Like signing a little-noted treaty of amity and cooperation with ASEAN that opened the door to permanent representation and ultimately elevated a forum for engaging on high-stakes issues like the South China Sea. We've encouraged India's "Look East" policy as a way to weave another big democracy into the fabric of the Asia Pacific. We've used trade negotiations over the Trans-Pacific Partnership to find common ground with a former adversary in Vietnam. And the list goes on. Our effort has encompassed all the levers of powers and more that I've both discussed and that we have utilized.

And you can ask yourself: How could we approach an issue as thorny and dangerous as territorial disputes in the South China Sea without a deep understanding of energy politics, subtle multilateral diplomacy, smart economic statecraft, and a firm adherence to universal norms?

Or think about Burma. Supporting the historic opening there took a blend of economic, diplomatic, and political tools. The country's leaders wanted the benefits of rejoining the global economy. They wanted to more fully participate in the region's multilateral institutions and to no longer be an international pariah. So we needed to engage with them on many fronts to make that happen, pressing for the release of political prisoners and additional reforms while also boosting investment and upgrading our diplomatic relations.

Then there's China. Navigating this relationship is uniquely consequential, because how we deal with one another will define so much of our common future. It is also uniquely complex, because—as I have said on many occasions, and as I have had very high-level Chinese leaders quote back to me—we are trying to write a new answer to the age-old question of what happens when an established power and a rising power meet.

To make this work, we really do have to be able to use every lever at our disposal all the time. So we expanded our high-level engagement through the Strategic & Economic Dialogue to cover both traditional strategic issues like North Korea and maritime security, and also emerging challenges like climate change, cyber security, intellectual property concerns, as well as human rights.

Now, this approach was put to the test last May when we had to keep a summit meeting of the dialogue on track while also addressing a crisis over the fate of a blind human rights dissident who had sought refuge in our American Embassy. Not so long ago, such an incident might very well have scuttled the talks. But we have through intense effort, confidence building, we have built enough breadth and resilience into the relationship to be able to defend our values and promote our interests at the same time.

We passed that test, but there will be others. The Pacific is big enough for all of us, and we will continue to welcome China's rise—if it chooses to play a constructive role in the region. For both of us, the future of this relationship depends on our ability to engage across all these issues at once.

That's true as well for another complicated and important region: the Middle East and North Africa.

I've talked at length recently about our strategy in this region, so let me just say this. There has been progress: American soldiers have come home from Iraq. People are electing their leaders for the first time in generations, or ever, in Egypt, Tunisia, and Libya. The United States and our partners built a broad coalition to stop [Lybian dictator Muammar] Qadhafi from massacring his people. And a ceasefire is holding in Gaza. All good things. But not nearly enough.

Ongoing turmoil in Egypt and Libya point to the difficulties of unifying fractured countries and building credible democratic institutions. The impasse between Israel and the Palestinians shows little sign of easing. In Syria, the Assad regime continues to slaughter its people and incite intercommunal conflict. Iran is pursuing its nuclear ambitions and sponsoring violent extremists across the globe. And we continue to face real terrorist threats from Yemen and North Africa.

So I will not stand here and pretend that the United States has all the solutions to these problems. We do not. But we are clear about the future we seek for the region and its peoples. We want to see a region at peace with itself and the world—where people live in dignity, not dictatorships, where entrepreneurship thrives, not extremism. And there is no doubt that getting to that future will be difficult and will require every single tool in our toolkit.

Because you can't have true peace in the Middle East without addressing both the active conflicts and the underlying causes. You can't have true justice unless the rights of all citizens are respected, including women and minorities. You can't have the prosperity or opportunity that should be available unless there's a vibrant private sector and good governance.

And of this I'm sure: you can't have true stability and security unless leaders start leading; unless countries start opening their economies and societies, not shutting off the internet or undermining democracy; investing in their people's creativity, not fomenting their rage; building schools, not burning them. There is no dignity in that and there is no future in it either.

Now, there is no question that everything I've discussed and all that I left off this set of remarks adds up to a very big challenge that requires America to adapt to these new realities of global power and influence in order to maintain our leadership. But this is also an enormous opportunity. The United States is uniquely positioned in this changing landscape.

The things that make us who we are as a nation—our openness and innovation, our diversity, our devotion to human rights and democracy—are beautifully matched to the demands of this era and this interdependent world. So as we look to the next four years and beyond, we have to keep pushing forward on this agenda, consolidate our engagement in the Asia Pacific without taking our eyes off the Middle East and North Africa; keep working to curb the spread of deadly weapons, especially in Iran and North Korea; effectively manage the end of our combat mission in Afghanistan without losing focus on al-Qaida and its affiliates; pursue a far-ranging economic agenda that sweeps from Asia to Latin America to Europe.

And keep looking for the next Burmas. They're not yet at a position where we can all applaud, but which has begun a process of opening. Capitalize on our domestic energy renewal and intensify our efforts on climate change, and then take on emerging issues like cyber security, not just across the government but across our society.

You know why we have to do all of this? Because we are the indispensable nation. We are the force for progress, prosperity and peace. And because we have to get it right for ourselves. Leadership is not a birthright. It has to be earned by each new generation. The reservoirs of goodwill we built around the world during the 20th century will not last forever. In fact, in some places, they are already dangerously depleted. New generations of young people do not remember GIs liberating their countries or Americans saving millions of lives from hunger and disease. We need to introduce ourselves to them anew, and one of the ways we do that is by looking at and focusing on and working on those issues that matter most to their lives and futures.

So because the United States is still the only country that has the reach and resolve to rally disparate nations and peoples together to solve problems on a global scale, we cannot shirk that responsibility. Our ability to convene and connect is unparalleled, and so is our ability to act alone whenever necessary.

So when I say we are truly the indispensible nation, it's not meant as a boast or an empty slogan. It's a recognition of our role and our responsibilities. That's why all the declinists [those who argue the United States is a declining power] are dead wrong. It's why the United States must and will continue to lead in this century even as we lead in new ways. And we know leadership has its costs. We know it comes with risks and can require great sacrifice. We've seen that painfully again in recent months. But leadership is also an honor, one that Chris Stevens and his colleagues in Benghazi embodied. And we must always strive to be worthy of that honor.

That sacred charge has been my north star every day that I've served as Secretary of State. And it's been an enormous privilege to lead to the men and women of the State Department and USAID. Nearly 70,000 serving here in Washington and in more than 270 posts around the world. They get up and go to work every day, often in frustrating, difficult, and dangerous circumstances, because they believe, as we believe, that the United States is the most extraordinary force for peace and progress the world has ever known.

And so today, after four years in this job, traveling nearly a million miles and visiting 112 countries, my faith in our nation is even stronger, and my confidence in our future is as well. I know what it's like when that blue and white airplane emblazoned with the words "United States of America" touches down in some far-off capital and I get to feel the great honor and responsibility it is to represent the world's indispensable nation. I'm confident that my successor and his successors and all who serve in the position that I've been so privileged to hold will continue to lead in this century just as we did in the last—smartly, tirelessly, courageously—to make the world more peaceful, more safe, more prosperous, more free. And for that, I am very grateful.

Hillary Rodham Clinton is a former U.S. Secretary of State (2009–2013), U.S. senator from New York State (2001–2009), and first lady of the United States (1993–2001). She holds a law degree from Yale University.

Ted Galen Carpenter    **NO**

Delusions of Indispensability

One striking feature of foreign-policy discussions in the United States is the widespread assumption that this country is the "indispensable nation" in the international system. Historian James Chace and Clinton presidential aide Sidney Blumenthal apparently coined the term in 1996 to capture the essence of Bill Clinton's liberal-internationalist vision of the post–Cold War world, but it is a term that conservatives and moderates as well as liberals have used frequently since then. In his 2012 State of the Union address, Barack Obama asserted that "America remains the one indispensable nation in world affairs—and as long as I'm President, I intend to keep it that way."

Only a handful of iconoclasts in the foreign-policy community—and even fewer mavericks in the political arena—dare to challenge the conventional wisdom. That is unfortunate, because the notion of the United States as the indispensable nation is not only dubious, but it also entrenches a counterproductive security strategy. It is a blueprint for strategic overextension and, ultimately, a failed paradigm.

The term "leadership" itself is often a euphemism for those who see the United States as the indispensable nation. They usually mean America as the de facto global hegemon, and some are occasionally candid enough to use that word. Mitt Romney succinctly expressed the concept when he asserted that "America is not destined to be one of several equally balanced global powers." Discussing the U.S. role in East Asia, American Enterprise Institute scholar Daniel Blumenthal warned of dire consequences if the United States no longer played "the role of benign hegemon in Asia."

Although some pundits and policy experts suggest that U.S. leaders should encourage other "cooperative" countries to have a greater voice and play a larger role in collaborative enterprises, even such proponents of multilateralism tend to become anxious if the United States is not clearly in charge on important matters. The neoconservative faction in the U.S. policy community does not even pretend to favor genuine multilateralism. Their preferred strategy is one in which the United States either acts unilaterally—often with a tinge of contempt for the views of other countries—or acts as the undisputed leader of a coalition, as during the Iraq War.

Proponents of the indispensable-nation thesis all agree that it would be calamitous for Washington to step back from its current global role. Such a move, in their view, would damage crucial U.S. interests, as well as the overall peace and prosperity of the world. They differ among themselves, though, on just what form that calamity would take.

For most, the primary danger they foresee is that chaos would ensue if the United States did not exercise robust global leadership. Writing in 2000, William Kristol, founder and editor of the *Weekly Standard*, and Robert Kagan, at the time a scholar at the Carnegie Endowment for International Peace, warned that the greatest danger in the twenty-first century was that the United States, "the world's dominant power on whom the maintenance of international peace and the support of liberal democratic principles depends," would "shrink its responsibilities and—in a fit of absentmindedness, or parsimony, or indifference—allow the international order that it created and sustains to collapse." Blumenthal agrees that such a terrible fate would certainly befall Asia, insisting that without U.S. hegemony, "chaos would ensue. No one would lead efforts to further build upon an economically vital region, stem proliferation, or keep great power peace."

Stuart Gottlieb, a former foreign-policy adviser to two senior Democratic senators, insists that history confirms that point on a global basis. He argues:

> Over the past century, each of America's attempts to reduce its role in the world was met by rising global threats, eventually requiring a major U.S. re-engagement. . . . In each case, hopes were soon dashed by global challengers who took advantage of America's effort to draw back from the world

stage—Germany and Japan in the 1930s, the Soviet Union in the immediate post–World War II period and the Soviet Union again after Vietnam. In each case, the United States was forced back into a paramount global leadership role.

Although most proponents of continued U.S. dominance argue that global chaos would be the inescapable consequence, there is another concern, especially among hawkish conservatives. While they do fret about the possibility of planetary anarchy, their primary worry is somewhat different. Orthodox believers in America's indispensability assume that no other nation or combination of nations could fill the void Washington's retrenchment would create. But some advocates of U.S. preeminence disagree, believing that other major nations would move to fill such a leadership vacuum. And they fear that the most likely replacements are nations whose values and policies are hostile to America's interests.

When not writing pieces with the apocalyptic Kristol, Kagan hedges his bets between the two unsavory aftermaths of U.S. withdrawal:

> The present world order—characterized by an unprecedented number of democratic nations; a greater global prosperity, even with the current crisis, than the world has ever known; and a long peace among great powers—reflects American principles and preferences, and was built and preserved by American power in all its political, economic, and military dimensions. If American power declines, this world order will decline with it. It will be replaced by some other kind of order, reflecting the desires and the qualities of other world powers. Or perhaps it will simply collapse, as the European world order collapsed in the first half of the twentieth century.

Kagan's one certainty is that a world without U.S. dominance would be an unpleasant place. As he writes, "The belief, held by many, that even with diminished American power 'the underlying foundations of the liberal international order will survive and thrive,' as the political scientist G. John Ikenberry has argued, is a pleasant illusion." Not surprisingly, most of the suspicion about potential new hegemons is directed at two major powers that are not liberal democracies: Russia and China. The jaundiced American view of Moscow has diminished, given the demise of the Soviet Union and the decline of Russian political and military clout since the end of the Cold War. But that change is more modest than one might expect. Russophobes may not hate and fear Vladimir Putin

as much as they did the likes of Stalin, Khrushchev and Brezhnev, but an undercurrent of hostility remains. The shrill reactions of Bush and Obama administration officials (and many pundits) to Russia's reluctance to endorse harsher sanctions against Iran—and more recently, against Syria—is one manifestation. Staunch proponents of NATO [North Atlantic Treaty Organization] suggest privately, and sometimes even publicly, that Moscow might again seek to dominate Eastern and Central Europe in the absence of a robust, U.S.-led NATO.

The principal suspicions, though, are directed against a rising China. Princeton University's Aaron Friedberg, one of the more sensible and moderate neoconservatives, argues that China is already determined to displace the United States as East Asia's hegemon. His latest book, *A Contest for Supremacy*, presents that thesis emphatically. Cambridge University's Stefan Halper, a former official in both the Nixon and Reagan administrations, believes that trend is not confined to Asia. In his recent book *The Beijing Consensus*, Halper contends that China is presenting itself as a rival global model (one of authoritarian capitalism) to the democratic-capitalist model (the "Washington Consensus") that the United States guards and sustains. He argues further that Beijing has made considerable inroads in recent years.

It would be a mistake, though, to assume that devotees of the indispensable-nation thesis only fear that hostile, undemocratic nations would move to fill a regional or global leadership vacuum. U.S. policy makers in both the Clinton and George W. Bush administrations were also noticeably unenthusiastic about even another democratic nation—or a combination of democratic nations—supplanting Washington's leadership role. During the 1990s, two editions of the Pentagon's policy-planning guidance document for East Asia made veiled warnings that another nation might step forward—and not in a way consistent with American interests. Given China's modest economic capabilities and military weakness at the time, the Pentagon's concerns did not seem directed primarily at that country. Both the language in those documents and the strategic context suggested that the principal worry of the Department of Defense planners was that Japan might remilitarize and eventually eclipse U.S. power and influence in the region.

The U.S. attitude toward a greater—and especially a more independent—security role for the European Union and its leading members has not been much friendlier. Time and again, the Clinton and Bush administrations discouraged their European allies from being more assertive and proactive about the Continent's security needs. Members of the U.S. policy community viewed with uneasiness and suspicion any move that threatened the preeminence of NATO as Europe's primary security institution. This was not surprising. Washington not only has a chair at NATO's table, it occupies the chair at the head of the table. Conversely, there is no U.S. seat when the EU makes decisions. For Americans who relish Washington's dominance in transatlantic affairs, that absence of an official U.S. role is troubling enough on important economic issues. They deem such a development on security issues even more worrisome.

Another prominent feature of the indispensable-nation thesis is that its adherents adopt the "light-switch model" of U.S. engagement. In that version, there are only two positions: on and off. Many, seemingly most, proponents of U.S. preeminence do not recognize the existence of options between the current policy of promiscuous global interventionism and "isolationism." Following President Obama's second inaugural address, *Wall Street Journal* columnist Bret Stephens was most unhappy with the sections on foreign policy. The title of his column, "Obama's You're-On-Your-Own World," conveyed his thesis in a stark manner. Obama's worldview, Stephens asserted, constituted a species of isolationism. Such an indictment of Barack Obama—the leader who escalated the war in Afghanistan, involved the United States in the Libyan civil war, led the charge for harsher sanctions against Iran, Syria and North Korea, and is pursuing a strategic pivot toward East Asia in large part to contain China's power—would seem to strain credulity. But for the more zealous proponents of U.S. dominance such as Stephens, even rhetorical hints of modest retrenchment in portions of the world are reasons for alarm.

Adherence to the light-switch model reflects either intellectual rigidity or an effort to stifle discussion about a range of alternatives to the status quo. Even in the security realm there are numerous options between the United States as the global policeman—or what it has become over the past two decades, the global armed social worker—and refusing to take any action unless U.S. territory is under direct military assault. It is extraordinarily simplistic to imply that if Washington does not involve itself in civil wars in the Balkans, Central Asia or North Africa, that it would therefore automatically be unwilling or unable to respond to aggressive actions in arenas that are more important strategically and economically to genuine American interests. Indifference about what faction becomes dominant in Bosnia or Mali does not automatically signify indifference if China attempted to coerce Japan.

Selectivity is not merely an option when it comes to embarking on military interventions. It is imperative for a major power that wishes to preserve its strategic solvency. Otherwise, overextension and national exhaustion become increasing dangers. Over the past two decades, the United States has not suffered from a tendency to intervene in too few cases. Quite the contrary, it has shown a tendency to intervene in far too many conflicts. But many of the opinion leaders who stress the need for constant U.S. global leadership advocate even more frequent and far-ranging U.S. actions. Washington Post columnist Richard Cohen takes President Obama to task for not being more proactive against the Syrian government. Cohen argues further that a "furious sense of moral indignation" must return to U.S. foreign policy. Indeed, it should be "the centerpiece" of that policy.

His comments illustrate a worrisome absence of selectivity regarding military interventions among members of the indispensable-nation faction. There is always an abundance of brutal crackdowns, bloody insurrections and nasty civil wars around the world. If a sense of moral indignation, instead of a calculating assessment of the national interest, governs U.S. foreign policy, the United States will become involved in even more murky conflicts in which few if any tangible American interests are at stake. That is a blueprint for endless entanglements, a needless expenditure of national blood and treasure, and bitter, debilitating divisions among the American people. A country that has already sacrificed roughly 6,500 American lives and nearly $1.5 trillion in just the past decade pursuing nation-building chimeras in Iraq and Afghanistan should not be looking to launch similar crusades elsewhere.

Not only do disciples of the indispensable-nation doctrine seem to regard engagement as a binary light switch, they fail to distinguish between its various manifestations. The thesis that engagement can take different forms (diplomatic, military, economic and cultural) and that U.S. involvement in each form does not have to be at the same level of intensity is apparently a revolutionary notion bordering on apostasy. To those disciples, the security aspect dominates everything else. Mitt Romney warned that America must lead the world or the world will become a more dangerous place, "and liberty and prosperity would surely be among the first casualties." Among the dangers Kagan projects is "an unraveling of the international economic order," because, among other reasons, "trade routes and waterways ceased to be as secure, because the U.S. Navy was no longer able to defend them."

Proponents of an expansive U.S. posture repeatedly assert that a peaceful international system, which is the also the foundation of global prosperity, requires a hegemon. They most frequently cite Britain in the nineteenth century and the United States from the end of World War II to the present, although some even point to the Roman Empire as evidence for their thesis. In his book *The Case for Goliath*, Johns Hopkins University's Michael Mandelbaum even asserts that the United States performs many of the benevolent stabilizing functions that a world government would perform. That, in his view, has been enormously beneficial both for the United States and for the world.

Leaving aside the ultimate fate of the Roman Empire, or even the milder but still painful decline of Britain—which were in part consequences of the economic and security burdens those powers bore—the hegemonic model is hardly the only possible framework for a relatively stable and peaceful international system.

There are constructive alternatives to the stifling orthodoxy of the United States as the indispensable nation. That is especially true in the twenty-first century. Not only are there multiple major powers, but a majority of those powers share the democratic-capitalist values of the United States and are capable of defending and promoting those values. Moreover, even those great powers that represent a more authoritarian capitalist model, such as Russia and China, benefit heavily from the current system characterized by open trade and an absence of armed conflict among major powers. They are not likely to become aggressively revisionist states seeking to overturn the international order, nor are they likely to stand by idly while lesser powers in their respective regions create dangerous disruptions.

The most practical and appealing model is a consortium of powerful regional actors, with the United States serving as a first among equals. The opportunity for Washington to off-load some of its security responsibilities is most evident in Europe. Making the change to a more detached security strategy there would offer important benefits to the United States at a low level of risk. It made a reasonable amount of sense for Washington to assume primary responsibility for the security of democratic Europe in the aftermath of World War II. Western Europe was the most important strategic and economic prize of that era, and a powerful, expansionist Soviet Union eyed that prize. The Western European powers, traumatized and exhausted by World War II, were not in a good position to resist Moscow's power and blandishments. U.S. leadership was nearly inescapable, and it was warranted to protect and promote important American interests.

But even during the final decades of the Cold War, the U.S. security blanket unfortunately caused an excessive and unhealthy dependence on the part of democratic Europe. And with the demise of the Soviet Union, a policy based on U.S. dominance now reeks of obsolescence. Despite its recent financial struggles, the European Union collectively has both a population and an economy larger than those of the United States. And Russia, if it poses a threat at all, is a far less serious menace than was the Soviet Union. Yet U.S. leaders act as though the European Union nations are inherently incapable of managing Europe's security affairs. And for their part, the European allies are content to continue free riding on Washington's exertions, keeping their defense budgets at minimal levels and letting the United States take primary responsibility for security issues that affect Europe far more than America.

Even a modest increase in defense spending by the principal European powers would enable the EU to handle any security problems that are likely to arise in the region. In that sense, Washington's dominant role in dealing with the Balkan conflicts in the 1990s was not evidence of the continuing need for U.S. leadership, but rather underscored the negative consequences of having encouraged Europe's security dependence on the United States for so many decades. The reality is that the threat environment in Europe is quite benign. There are few plausible security threats, and the ones that might arise are on the scale of the Balkan spats—problems that the European powers should be able to handle without undue exertion. Washington can safely off-load responsibility for European security and stability to the countries directly involved. The United States is most certainly not indispensable to the Continent's security any longer.

Prospects in other regions are less definite, but there are still opportunities for Washington to reduce its military exposure and risks. The most important region to the United States, East Asia, presents a less encouraging picture than does Europe for off-loading security obligations, since there is no cohesive, multilateral organization comparable to the EU to undertake those responsibilities. Yet even in East Asia there are alternatives to U.S. hegemony, which has been in place since 1945.

Washington's dominance was born in an era in which there were no credible challengers. Although the USSR [Union of the Soviet Socialist Republics] had some ambitions in the western Pacific, its primary goals were elsewhere, largely in Eastern Europe and the emerging states of the Third World. China after the Chinese Revolution in 1949 was belligerent, but also weak and poor. Japan, utterly defeated in World War II and worried about Soviet and Chinese intentions, was content to maintain a pacifist image and rely heavily on the United States for defense. The rest of the region consisted of new, weak states arising out of rapidly decaying European colonial empires.

As in Europe, the situation today is totally different. Japan has the world's third largest economy, China is an emerging great power, and East Asia has an assortment of other significant economic and political players. It will be increasingly difficult for the United States, a nation thousands of miles away, to dominate a region with an ever-expanding roster of major powers.

Instead of frantically trying to prop up a slipping hegemony, U.S. policy makers must focus on helping to shape a new security environment. Among other steps, Washington should wean its principal allies in the region—especially Japan, South Korea and Australia—from their overreliance on U.S. defense guarantees. Not only should U.S. leaders make it clear that the United States intends to reduce its military presence, but they should emphasize that those allies now must take far greater responsibility for their own defense and the overall stability of the region.

The most likely outcome of such a policy shift would be the emergence of an approximate balance of power in East Asia. China would be the single strongest country, but if Japan, South Korea, and other actors such as Vietnam and Indonesia take the actions necessary to protect their own interests, Beijing will fall far short of having enough power to become the new hegemon. A balance-of-power system would be somewhat less stable than the current arrangement, but it would likely be sufficient to protect crucial American interests. And it may be Washington's only realistic option over the medium and long term. Clinging to an increasingly unsustainable hegemony is not a realistic strategy.

Off-loading security responsibilities in other regions needs to be assessed on a case-by-case basis. In most instances, adverse developments in those regions affect other major powers more than they do the United States. It is a bit bizarre, for example, that Washington should take more responsibility for developments in the Middle East than do such NATO allies as Germany, France, Italy and Turkey. Or that Washington is more concerned about troubles in South and Southeast Asia than are major powers such as India and Indonesia. But other relevant actors have not had to step forward to deal with unpleasant developments that might undermine regional stability, because the self-proclaimed indispensable nation has usually taken on the responsibility. That is not sustainable.

In the all-too-rare instances in which the United States did not seek to take care of problems that mattered more to other powers, those countries did not inevitably sit back and watch the situation deteriorate. One example occurred when conflict broke out between rival factions in Albania in the late 1990s and Washington declined to lead yet another intervention in the Balkans. Faced with the U.S. refusal, Italy and Greece organized and led an ad hoc European military coalition that restored order before the turmoil could intensify and spread beyond Albania's borders. Various foreign-policy experts have presented detailed cases for options that would reduce the extent—and hence the costs and risks—of America's security role. Boston University's Andrew Bacevich, Texas A&M's Christopher Layne and the Cato Institute's Christopher Preble are just some of the more prominent analysts who chart a course between the extremes of the current policy and Fortress America. All of them, to one extent or another, make the case for off-loading at least some of Washington's security commitments onto other capable powers and adopting a new, more restrained posture of "offshore balancing."

The notion that the United States is the indispensable nation is a conceit bordering on narcissism. It had some validity during an era of stark bipolarity when a weak, demoralized democratic West had to depend on American power to protect the liberty and prosperity of the non-Communist world from Soviet coercion. But the world has been multipolar economically for decades, and it has become increasingly multipolar diplomatically and politically in recent years. Yet so much of the American political and foreign-policy communities embrace a security role—and an overall leadership role—for the United States that was born in the era of bipolarity and perpetuated during what Charles Krauthammer described as the "unipolar moment" following the collapse of the Soviet empire.

That moment is gone, and that is not the world we live in today. The United States needs a security strategy appropriate for a world of ever-increasing multipolarity. Very few critics of U.S. hegemony advocate an abandonment of all of America's security commitments. But an aggressive pruning of those commitments is overdue. It is well past time for the EU to assume primary responsibility for Europe's security and for Japan to emerge as a normal great power with appropriate ambitions and responsibilities in East Asia. It is also past time for smaller U.S. allies, such as South Korea and Australia, to increase their defense spending and take more responsibility for their own defense. While the off-loading of Washington's obligations needs to be a gradual process, it also needs to begin immediately and to proceed at a brisk pace. And Washington ought to make it clear to all parties concerned that it is entirely out of the business of nation building.

Those who desperately try to preserve a status quo with America as the indispensable nation risk an unpleasant outcome. A country with America's financial woes will find it increasingly onerous to carry out its vast global-security commitments. That raises the prospect of a sudden, wrenching adjustment at some point when the United States simply cannot bear those burdens any longer. That is what happened to Britain after World War II, when London had no choice but to abandon most of its obligations in Africa, Asia and the Mediterranean. The speed and extent of the British move created or exacerbated numerous power vacuums. It is far better for the United States to preside over an orderly transition to an international system in which Washington plays the role of first among equals, rather than clinging to a slipping hegemony until it is forced to give way.

TED GALEN CARPENTER, a senior fellow at the Cato Institute and a contributing editor to *The National Interest*, is the author of numerous books on international affairs, including *Smart Power: Toward a Prudent Foreign Policy for America* (Cato Institute, 2008). He received his Ph.D. in U.S. diplomatic history from the University of Texas.

EXPLORING THE ISSUE

Should the United States Seek to Remain the "Indispensable" Country?

Critical Thinking and Reflection

1. Is there really any such thing as American national interests that can be objectively identified or is every so-called interest no more than the subjective view of those who advocate it? If there are objective interests—objective or subjective—on which all or nearly all American can agree, what are they?
2. Those who favor the United States playing a wide and powerful role in world affairs predict that without the U.S. presence the world would become both much more unstable and dangerous and also that such values as human rights would decline. Some suggest, for example, that nuclear proliferation will explode, with countries like Japan, Germany, South Korea, Iran, Brazil, and Argentina acquiring nuclear weapons to provide the protection that the U.S. "nuclear umbrella" once did. Of course, no one knows for sure. Is the United States' "retreat from power" worth the risk of a world that resembles the unstable, horrific first half of the twentieth century?
3. To what degree should U.S. foreign policy be driven by "narrow" national interests such as protecting national territory and the lives of Americans, on the one hand, and "broader" national interests like promoting democracy, furthering and protecting human rights, and promoting an environmentally sustainable world?

Is There Common Ground?

The answer to this question is yes, there is. What that common ground might be is less certain. There is something of a continuum of roles that the United States could seek to play in the world. At one end, the United States could seek to be a unipolar power that dominates the global international system. Isolationism is at the other end of the continuum. This policy would have the United States strictly limit its global involvement, particularly military activity, to cases where the United States, its territories, its forces, and perhaps its citizens, at least in substantial numbers, have been attacked or are under immediate threat. Few today, including Hillary Clinton and ted Galen Carpenter, argue for either extreme. There are, however, myriad points in between. The difficulty, of course, is where, when, and how to be involved.

Determining that begins with the evaluation of American national interests, as asked in the section on critical thinking questions. Enumerating them is not enough. Interests also need to be put into some order of priority or at least grouped into categories, such as vital, important, and peripheral and perhaps long- and short-term. Since pursing the national interest is seldom cost free, it is also necessary to have some idea how much defending or advancing each interest is worth. This may be the most difficult part. Determining how much to spend each year

on the military is, for instance, relatively easy compared to deciding how many American soldiers' lives it is worth to, say, prevent Iran from getting nuclear arms, defend the U.S.-NATO ally Latvia if it is invaded by the neighboring Russia, or halt a genocide slaughter in Rwanda, Bosnia, Syria, or one of the too many other places one has occurred. It is impossible to make such calculations, you might object, but in reality such decisions are at least implicitly made each time the military is used.

Create Central

www.mhhe.com/createcentral

Additional Resources

Brooks, Stephen G., Ikenberry, John, and Wohlforth, William C. (Winter 2012/13). "Don't Come Home, America: The Case against Retrenchment," *International Security* (37/3: 7–51)

This article delves deeply into both the theory and realities of the distribution of global power. It argues for the United States trying to pay the indispensable role.

Friedman, Benjamin H., Green, Brendan Rittenhouse, and Logan, Justin (Fall 2003). "Correspondence:

Debating American Engagement: The Future of U.S. Grand Strategy," *International Security* (38/2: 181–199)

This article replies to and disagrees with the Brooks, Ikenberry, and Wohlforth article above.

Shultz, George P. (Fall 2013). "Still the Essential Nation," *Hoover Digest* (4:134–140)

Shultz, who served as secretary of state in the Reagan administration, advocates the United States staying fully engaged in world politics.

Internet References . . .

Quadrennial Defense Review, 2014

Issued by the U.S. Department of Defense, the QDR is a review required by Congress to be conducted every four years that analyzes threats to U.S. national security and the U.S. military capability to respond. Annual updates are published.

**http://www.defense.gov/pubs/2014_Quadrennial_
Defense_Review.pdf**

National Security Policy, 2010

The latest of these reports, issued by the Obama administration in May 2010, lays out a strategic approach for advancing American interests, including the security of the American people, a growing U.S. economy, support for our values, and an international order that can address twenty-first-century challenges.

**http://www.whitehouse.gov/sites/default/files/rss_
viewer/national_security_strategy.pdf**

Global Firepower—2014 World Military Strength Rankings

Military power is only one aspect of national power, and quantitative rankings do tell the whole story of military power. Still, this data site gives good insights into U.S. power and the huge edge the United States has over any other country.

www.globalfirepower.com

World Fact Book

An excellent accumulation of facts about all the countries of world. Also allows comparisons among countries in many power-relevant categories such as military spending, gross domestic product, and trade. Revised annually.

**https://www.cia.gov/library/publications/the-world
-factbook/**

Selected, Edited, and with Issue Framing Material by:
John T. Rourke, *University of Connecticut, Storrs*

ISSUE

Is Russia Being Demonized for Pursuing Its National Interests?

YES: Vladimir Putin, from "Address by President Vladimir Putin at the Conference of Russian Ambassadors and Permanent Representatives," Permanent Mission to the Russian Federation to the European Union, Moscow (2014)

NO: North Atlantic Treaty Organization, from "Russia's Accusations: Setting the Record Straight," Factsheet Issued by NATO (April, 2014) and Updated (July 2014), on the NATO Website at: http://www.nato.int/cps/en /natolive/topics_111767.htm (2014)

Learning Outcomes

After reading this issue, you will be able to:

- Relate the grounds on which President Putin claims that Russia is being unfairly criticized.
- Discuss the changes since the collapse of the Soviet Union in the state of democracy in Russia.
- Explain the events leading up to Russia's crisis with Ukraine beginning with the seizure of the Crimea.
- Outline the position and reactions of the United States and Western Europe to the crisis.

ISSUE SUMMARY

YES: Vladimir Putin, the president of Russia, tells a gathering of Russia's diplomats that their country's response to events in Ukraine represents a legitimate reaction to threats to ethnic Russians living in Ukraine and to Russia national security interests. He lays the blame on Ukrainian nationalism and on a NATO-led effort to bring threatening forces ever closer to Russia's borders.

NO: A report issued by NATO asserts that it is Russia's aggression against Ukraine that has led to a crisis. NATO further argues that to divert attention away from its actions, Russia has leveled a series of false accusations against NATO.

Russia experienced two momentous revolutions during the twentieth century. The first began in March 1917 with the events that soon led to the fall of the czar's government and the establishment of the Union of the Soviet Socialist Republics (USSR), with Russia one of its 15 republics.

The second great revolution occurred in 1991 when the USSR collapsed and its 15 republics became independent countries. Of these former Soviet republics (FSRs), Russia is by far the most important in terms of territory, population, resources, and power. Russia retained the bulk of the Soviet Union's military assets including its nuclear weapons. Soon, however, the exhilaration of Russia's renewed independence faded amid myriad problems. The Russian economy declined sharply, and that led to a decline in Russia's military capabilities.

For good or ill, Russia's fortunes began when its second president, Vladimir Putin, was elected in 2000. He brought a level of stability and rejuvenation to Russia. Russia's economy has improved substantially, and Russia's military forces have regained at least some of their former capacity. Russia is certainly not the superpower the Soviet Union was, but Russia is once again a major power and far overshadows the strength of the other former FSRs, including Ukraine, and the countries of Eastern Europe that were once Soviet client states and allies.

Putin served two terms as president (2000–2008), but was ineligible to run for a third term. In what was little more than a sham transfer of power, Putin's prime minister, Dmitry Medvedev served as president from 2008 to 2012. Then Putin, eligible once again, ran for president and was elected to a third term in a five-way contest with 64 percent of the vote. His nearest opponent managed only 17 percent of the vote. Since then, Putin's popularity among Russians has continued to grow and reached a record high 87 percent positive rating in August 2014.

There have been two important trends under Putin. First, Russia has become steadily less democratic. Most of the independent news media is gone, local authorities have lost much of their power, and opposition movement and leaders are increasingly repressed and individuals are sometimes jailed. Surely Russia is not a totalitarian dictator in the mold of Vladimir Lenin or Josef Stalin, but neither can Putin be characterized as a democratic leader.

Second, Russia has moved with increasing emphasis to rebuild its power and to reassert its international and, in particular, its regional power. Putin and most Russians believe that their country should have the important place in global affairs, just as the USSR, and before it, czarist Russia, had.

This reassertion of power has caused or worsened a variety of issues that divide Russia from the United States and its NATO allies. For example, Moscow believes that it is threatened by the U.S. drive to deploy parts of a ballistic missile defense system to countries near Russia's borders. Russians also see a threat in the expansion of NATO to now include most of the Soviet Union's former East European allies and even three FSRs (Estonia, Latvia, and Lithuania). Russians are also alarmed by the expansion of the European Union (EU) to Russia's borders. Most East European countries are now EU members, including Estonia, Latvia, and Lithuania. Indeed, one of the issues that worried Russia about events in Ukraine was a strong movement in that FSR to also join the EU and even NATO.

The crisis centered in Ukraine began in late 2013 when protests broke out against the effort of Ukrainian President Viktor Yanukovych to reverse his country's movement toward EU membership. He was forced from power, and in May 2014 a new president, Petro Poroshenko, was elected. He had campaigned on a pro-European Union platform.

That was soon followed by Russia seizing the Crimea. It was part of Ukraine but also the site of Russia's major naval base on the Black Sea, with its access to the Mediterranean Sea. The seizure was generally portrayed in the West a clear aggression, but there are mitigating factors. One is that Crimea was part of Russian territory beginning in 1783 and only became part of Ukraine in 1954 in a shift involved with arcane Soviet internal politics. Second, the majority of people in Crimea were not and never had been Ukrainians. According to the 2001 Ukrainian census, Russians made up 58 percent of the Crimea's population, followed by Ukrainians (24%) and Crimean Tatars (12%). Adding to the unrest, ethnic Russians make up about 30 percent of the population in Ukraine's four most eastern provinces. These ethnic Russians, at least with some encouragement by Moscow, argued they feared Ukrainian oppression and launched a secessionist movement.

Reflected by the NO selection, the NATO report, the overwhelming response to these events in the United States and Europe has been to condemn Russia for aggression. But Vladimir Putin and most Russians see the matter in a very different light, as evident in the YES selection. For Russians the issue is what seems to be the encroachment right to Russia's borders of their former enemies in NATO and what they see as an associated EU. This move from Russians' perspective is an effort to keep Russia from resuming its rightful place as a world standing and role as the dominant country in its region, just as the United States dominates its sphere of influence in the Western Hemisphere and China is seeking greater sway in Southeast Asia and elsewhere near its borders.

YES ⤶

Vladimir Putin

Address by President Vladimir Putin at the Conference of Russian Ambassadors and Permanent Representatives

Meetings with the diplomatic corps [and Russia's president] have become a tradition. We need this direct conversation to make an overall assessment of the situation in the world, to set current and long-term foreign policy objectives and on that basis to more effectively coordinate the work of our missions abroad.

I would like to begin by saying that the Foreign Ministry and our embassies are under a lot of pressure; we see this, we are aware of this, but this pressure will not be reduced. It will only increase, just as the requirement to show efficiency, precision and flexibility in our actions to ensure Russia's national interests.

You know how dynamic and unpredictable international developments may sometimes be. They seem to be pressed together and unfortunately are not all of a positive nature. The potential for conflict is growing in the world, old contradictions are growing ever more acute and new ones are being provoked. We come across such developments, often unexpectedly, and we observe with regret that international law is not working, the most basic norms of decency are not complied with and the principle of all-permissiveness is gaining the upper hand.

We are observing this in Ukraine as well. We need to understand clearly that the events provoked in Ukraine are the concentrated outcome of the notorious deterrence policy. As you may know, its roots go deep into history and it is clear that unfortunately, this policy did not end with the end of the Cold War.

In Ukraine, as you may have seen, at threat were our compatriots, Russian people and people of other nationalities, their language, history, culture and legal rights, guaranteed, by the way, by European conventions. When I speak of Russians and Russian-speaking citizens I am referring to those people who consider themselves part of the broad Russian community; they may not necessarily be ethnic Russians, but they consider themselves Russian people.

What did our partners expect from us as the developments in Ukraine unfolded? We clearly had no right to abandon the residents of Crimea and Sevastopol to the mercy of nationalist and radical militants; we could not allow our access to the Black Sea to be significantly limited; we could not allow NATO [North Atlantic Treaty Organization] forces to eventually come to the land of Crimea and Sevastopol, the land of Russian military glory, and cardinally change the balance of forces in the Black Sea area. This would mean giving up practically everything that Russia had fought for since the times of Peter the Great [czar of Russia, 1682–1725], or maybe even earlier—historians should know.

I would like to make it clear to all: our country will continue to actively defend the rights of Russians, our compatriots abroad, using the entire range of available means—from political and economic to operations under international humanitarian law and the right of self-defense.

I would like to stress that what happened in Ukraine was the climax of the negative tendencies in international affairs that had been building up for years. We have long been warning about this, and unfortunately, our predictions came true.

You know about the latest efforts to restore, to maintain peace in Ukraine. Foreign Ministry staff and the Minister himself took an active part in this. You know about the numerous telephone conversations we had on this subject.

Unfortunately, [Ukraine's President Petro] Poroshenko has resolved to resume military action, and we failed—when I say "we," I mean my colleagues in Europe and myself—we

Address by President Vladimir Putin at the Conference of Russian Ambassadors and Permanent Representatives, Permanent Mission to the Russian Federation to the European Union, April 2014 (updated July 2014).

failed to convince him that the road to a secure, stable and inviolable peace cannot lie through war.

So far Mr. Poroshenko was not directly linked to the orders to begin military action, and only now did he take full responsibility, and not only military, but political as well, which is much more important.

We also failed to agree to make public the statement approved by the foreign ministers of Germany, France, Russia and Ukraine on the need to maintain peace and search for mutually acceptable solutions.

I would like to draw your attention to the fact that after the ceasefire was declared, no substantive, as you say, negotiations on the settlement of the situation ever began. Virtually, a disarmament ultimatum was given. However, even the ceasefire was not bad overall, though not enough to settle the situation on a long-term basis in a way that would be acceptable to all the people living in the country, including those in its southeast.

A constitution was made public, but it was never discussed. Even within Ukrainian society there is a discussion of whether it is good or bad, but nobody definitely ever discussed it with the east.

Of course, everything that is going on in Ukraine is the internal affair of the Ukrainian state. It pains us to see people dying, especially civilians. As you may know, the number of refugees in the Russian Federation is growing. We will of course provide assistance to all those who need it. However, killing journalists is unacceptable. I reminded the Ukrainian president of this yesterday yet again.

In my view, we are observing a focused effort to liquidate all media representatives. This applies to both Russian and foreign journalists. Who could be afraid of fair reporting? Probably those who are committing crimes. We strongly hope that the Ukrainian authorities act on their promises to carefully investigate the crimes.

More new hotspots are appearing on the world map. There is a deficit of security in Europe, in the Middle East, South-East Asia, in the Asia-Pacific region and in Africa. The global economic, financial and trade systems are becoming unbalanced, and moral and spiritual values are being washed out.

There is hardly any doubt that the unipolar world order did not come to be. Peoples and countries are raising their voices in favor of self-determination and civilizational and cultural identity, which conflicts with the attempts by certain countries to maintain their domination in the military sphere, politics, finance, economy and ideology.

I know this has no direct bearing on us, however what is being done to the French banks can cause nothing but indignation in Europe in general and here as well. We are aware of the pressure our American partners are putting on France to force it not to supply Mistrals to Russia. We even know that they hinted that if France does not deliver the Mistrals, the sanctions will be quietly lifted from their banks, or at least they will be significantly minimized. [The Mistral is a French large amphibious assault ship capable of launching helicopters. Similar ships are members of the Mistral class, and France agreed in 2010 to build two for Russia for $1.7 billion. The first is due in 2015, the second in 2017.]

What is this if not blackmail? Is this the right way to act on the international arena? Besides, when we speak of sanctions, we always assume that sanctions are applied pursuant to Article 7 of the UN Charter. Otherwise, these are not sanctions in the true legal sense of the word, but something different, some other unilateral policy instrument.

In the past 20 years, our partners have been trying to convince Russia of their good intentions, their readiness to jointly develop strategic cooperation. However, at the same time they kept expanding NATO, extending the area under their military and political control ever closer to our borders. And when we rightfully asked: "Don't you find it possible and necessary to discuss this with us?" they said: "No, this is none of your business." Those who continue insisting on their exclusivity strongly dislike Russia's independent policy. The events in Ukraine prove this. They also prove that a model of relations full of double standards does not work with Russia.

Nevertheless, I hope pragmatism will eventually prevail. We need to get rid of ambitions, of attempts to establish a "world barracks" and arrange everybody by rank, or to impose single rules of behavior and life, and to finally begin building relations based on equality, mutual respect and concern for mutual interests. It is time we admit each other's right to be different, the right of every country to live its own life rather than to be told what to do by someone else.

Colleagues, in its foreign policy Russia has been consistently proceeding from the notion that solutions to global and regional conflicts should be sought not through confrontation, but through cooperation and compromise. We advocate the supremacy of international law while supporting the UN's leading role.

International law should be mandatory for all and should not be applied selectively to serve the interests of individual select countries or groups of states, and most importantly, it should be interpreted consistently. It is impossible to interpret it in one way today, and in a different way tomorrow to match the political goals of the day.

World development cannot be unified. However, we can look for common issues, see each other as partners rather than competitors, and establish cooperation between states, their associations and integration structures.

These are the principles we were guided by in the past, and they continue to guide us now as we promote integration within the CIS. Strengthening close friendly ties and developing mutually advantageous economic cooperation with our neighbors is the key strategic priority of Russia's long-term foreign policy. [The CIS, or Commonwealth of Independent States, is a loose cooperation agreement created after the collapse of the Soviet Union in 1991 by Russia and all the other 15 former Soviet republics except Latvia, Lithuania, Estonia, and Georgia.]

The driving force behind Eurasian integration is the trio of Russia, Belarus and Kazakhstan [all former Soviet republics, FSRs]. The Agreement on the Eurasian Economic Union, signed in Astana [the capital of Kazakhstan] on May 29 [2014], symbolizes a qualitatively new step in our relations. A powerful center of economic development that attracts business and investors, a common market is being formed in Eurasia. That is why our CIS partners show a strong interest in this union. I hope that very soon, Armenia will become a full-fledged member of this union. Negotiations with [FSR] Kyrgyzstan are at an advanced stage. We are open to other Commonwealth states as well.

As we promote the Eurasian integration project, we are in no way trying to separate ourselves from the rest of the world; we are ready to consider prospects for creating free trade zones both with individual states and with regional associations and unions, primarily the European Union, of course.

Europe is our natural and most significant trade and economic partner. We strive to find new opportunities to expand our business cooperation, to open up new prospects for mutual investment and to lift trade barriers. This requires an upgrade of the legal contractual base of our cooperation and the stability and predictability of ties, primarily in such strategically important areas as energy. Stability on the entire territory of Eurasia and sustainable development of the EU economies and Russia depend on well-coordinated cooperation based on consideration for mutual interests.

We have always held high our reputation of a reliable supplier of energy resources and invested in the development of gas infrastructure. Together with European companies, as you may know, we have built a new gas transportation system called Nord Stream under the Baltic Sea. Despite certain difficulties, we will promote the South Stream project, especially since ever more European politicians and businessmen are coming to understand that someone simply wants to use Europe in their own interests, that it is becoming a hostage of someone's near-sighted ideologized approaches.

If we return to Ukraine, the violation by Ukraine of its commitments regarding the purchase of our natural gas has become a common problem. Kiev refuses to pay on its debt. This is absolutely unacceptable. They have not paid for November–December of last year [2013], though there were no arguments whatsoever then.

Our partners are using blatant blackmail—this is what it is. They demand an ungrounded reduction of prices on our goods, though the agreement was signed in 2009, and the parties complied with it in good faith. Now, as you may know, the court in Kiev [Ukraine's capital] has lifted all accusations against Ukraine's former Prime Minister Tymoshenko, who signed the contract. Thus, the Kiev court authorities admit that they have done everything right not only by international law, but by Ukrainian law as well. But they do not wish to comply, or to pay for the product already received.

As of June 16, as you may know, we have transferred Ukraine to a pre-payment system, so they will get exactly the amount of gas they pay for. Today they do not pay; therefore, they are not getting anything—only in the so-called reverse mode. We know all about this reverse mode: it is a fake; there is no reverse mode. How can you supply gas two ways along the same pipeline? One does not have to be a gas transportation expert to understand that this is impossible. They are playing tricks with some of their partners: in fact, they are getting our gas and paying some western partners in Europe who are not receiving their volume. We are quire aware of this.

We are not taking any action at this point only because we do not want the situation to deteriorate. However, everyone should draw the proper conclusions from the situation. The main thing is that honest gas consumers and suppliers should not suffer from the actions of Ukrainian politicians and bureaucrats.

Generally, all of us—Ukraine, our European partners, and we—should seriously consider how to reduce the probability of any type of political or economic risks or force majeure situations on the continent.

In this connection, I would like to remind you that in August 2015 we will be marking 40 years of the Final Act of the Conference on Security and Cooperation in Europe [CSCE]. This anniversary is a good reason not only to turn to the basic principles of cooperation on the continent that were laid back in 1975, but also to jointly make them work, to help them take root in practical European politics. [The CSCE created the Organization for Security and Cooperation in Europe, an organization with 57 member countries, including Russia and the United States, founded to promote peace, primarily in Europe.]

We have to work consistently to rule out any unconstitutional coups in Europe, any interference in the internal affairs of sovereign states, the use of blackmail or threats in international relations or the support of radical and neo-Nazi forces.

All of us in Europe need a sort of safety net to make sure that Iraqi, Libyan or Syrian—and unfortunately, I have to say also Ukrainian—precedents do not become contagious. This is especially dangerous for the post-Soviet area, because the states have yet not gained political or economic strength, they do not have a stable political system. It is very important that the constitutions of these states be treated with great care and respect.

Why is this important—and not only on the post-Soviet area, but all over Europe? Because even in those countries of Western and Eastern Europe where things seem to be going fine, there are quite a few hidden ethnic and social contradictions that may become acute any moment, may serve as ground for conflicts and extremism, and may be used by external forces to rock the social and political situation to achieve an illegitimate undemocratic change of power with all the negative consequences.

Firm guarantees of indivisible security, stability, respect for sovereignty and non-interference in each other's internal affairs should become the basis that we can use to build a common space for economic and humanitarian cooperation that would spread from the Atlantic Ocean to the Pacific Ocean—I already spoke of this as a single space from Lisbon to Vladivostok [a port in Russia on the Pacific Ocean].

I would like to ask the Foreign Ministry to draft a set of proposals in this respect, with special focus on the inadmissibility of any attempts to influence internal political processes from the outside. The job is to work the traditional principle of non-interference into the modern European realities and initiate a serious international discussion on the subject.

We also need to continue strengthening the eastern vector of our diplomacy, to more intensively use the impressive potential of the Asia-Pacific region in the interests of the further development of our country, primarily, of course, of Siberia and the Far East. We should continue to direct Russia's policy in Asia and the Pacific at maintaining the security of our eastern borders and at supporting peace and stability in the region. The coming leadership of Russia in the Shanghai Cooperation Organization, and the SCO and BRICS summits to be held in Ufa

in the summer of 2015 work to support this. [The SCO is an economic/military cooperation organization created in 2001 by China, Kazakhstan, Kyrgyzstan, Russia, Tajikistan, and Uzbekistan. BRICs denotes Brazil, Russia, China, and India and their economic cooperation talks.]

We need to strengthen overall partnership and strategic cooperation with the People's Republic of China. We can say that a strong Russian-Chinese connection has taken shape on the international arena. It is based on a coincidence of views on both global processes and key regional issues. It is of primary importance that Russian-Chinese friendship is not directed against anyone: we are not creating any military unions. On the contrary, this is an example of equal, respectful and productive cooperation between states in the 21st century.

We intend to further develop our relations with our traditional partners in this area of the world: with India and Vietnam, who are playing an ever-greater role in the world; with Japan and other countries, including the ASEAN [the ten-member country Association of South East Asian] states. We intend to further use the potential of the growing markets in Latin America and Africa and the great experience of political and humanitarian relations with the countries there.

Our contacts with the United States of America are of great importance for the whole world. We do not intend to shut down our relations with the USA. True, bilateral relations are not in their best shape, but—I would like to stress this—not through Russia's fault. We have always tried to be predictable partners and conduct our affairs on the basis of equality. However, in return, our lawful interests were often ignored.

Now over to various types of international meetings. If we are assigned the observer role without a decisive vote on key issues that are of vital importance to us, then such meetings are of little interest to us [this refers to NATO]. We should not sacrifice our vital interests just for the sake of being able to sit and observe. I hope our partners will eventually come to understand this obvious fact. So far, we have been hearing ultimatums or mentoring. Nevertheless, we are ready for dialogue, but I would like to stress that this should be an equal dialogue.

Colleagues, the complicated and unpredictable situation in the world places great demands on Russian diplomats' professional level. The Foreign Ministry's staff in Moscow and the Russian embassies abroad worked effectively and in coordinated fashion during the serious situation with Crimea and Ukraine, and I want to thank you for this. I particularly note the work done by the heads and staff of Russia's representative missions at the UN and other key international organizations.

We must continue working with just such energy and dignity, in a spirit of tact, restraint and sense of measure of course. Our position must be based on clear and unshakeable principles of international law and legal and historical justification, on truth, justice, and the strength of moral superiority.

VLADIMIR PUTIN is the president of Russia (2000–2008, 2012–present) and a former prime minister (2008–2012). He is a former officer of the national security agencies of the Soviet Union and Russia. Putin has a degree in international law from Leningrad State University.

North Atlantic Treaty Organization    **NO**

Russia's Accusations: Setting the Record Straight

Russia's aggression against Ukraine has led to Russia's international isolation, including NATO's suspension of all practical cooperation with Russia. To divert attention away from its actions, Russia has leveled a series of accusations against NATO which are based on misrepresentations of the facts and ignore the sustained effort that NATO has put into building a partnership with Russia. Russia has also made baseless attacks on the legitimacy of the Ukrainian authorities and has used force to seize part of Ukraine's territory. This document sets the record straight.

Russian Claims That the Ukrainian Authorities Are Illegitimate

Ukraine's President [Petro] Poroshenko was elected on 25 May with a clear majority in a vote which the OSCE [Organization for Security and Cooperation in Europe] characterized as showing the "clear resolve of the authorities to hold what was a genuine election largely in line with international commitments and with a respect for fundamental freedoms." The only areas where serious restrictions were reported were those controlled by separatists, who undertook "increasing attempts to derail the process." In other words, the president [of Ukraine] is legitimate, the actions of the separatists were not.

The current Ukrainian government was approved by an overwhelming majority in the Ukrainian parliament (371 votes out of 417 registered) on 27 February 2014, including members of the Party of Regions. That parliament was elected on 28 October 2012. The Russian Foreign Ministry at the time declared that the elections were held "peacefully, without any excesses and in line with generally accepted standards" and "confirmed Ukraine's commitment to democracy and the rule of law." The parliament which Russia called legitimate then can hardly be called illegitimate now.

Finally, Russian officials continue to allege that the Ukrainian parliament and government are dominated by "Nazis" and "fascists." However, in the presidential elections on May 25, the candidates whom Russia labeled as "fascists" received barely 1% of the votes. Ukraine's electorate clearly voted for unity and moderation, not separatism or extremism.

Russian Claims That NATO's Response Is Escalatory

Russian officials accuse NATO of escalating the crisis in Ukraine by reinforcing the defense of Allies in Eastern Europe. This is a striking display of double standards. It is Russia which is destabilizing Europe—not NATO.

Firstly, NATO's actions throughout the crisis have been proportionate to the situation, and defensive in nature. The Alliance has deployed additional aircraft to reinforce air policing missions, additional ships to the Baltic, Mediterranean and Black Seas, and additional troops to exercises on the territory of Eastern Allies.

All of these deployments are limited in scale and designed to reinforce defense. They have been prompted by the instability and unpredictability Russia has generated on our borders by its illegal invasion of a sovereign European country. NATO's actions cannot be presented as a potential offensive force. To describe them as such shows either ignorance or dishonesty. They are in line with NATO's international commitments, including the NATO-Russia Founding Act [a 1997 agreement to promote cooperation between Russia and NATO]. In the NATO-Russia Founding Act, NATO reiterates that:

> In the current and foreseeable security environment, the Alliance will carry out its collective defense and other missions by ensuring the necessary interoperability, integration, and capability

for reinforcement rather than by additional permanent stationing of substantial combat forces. Accordingly, it will have to rely on adequate infrastructure commensurate with the above tasks. In this context, reinforcement may take place, when necessary, in the event of defense against a threat of aggression and missions in support of peace consistent with the United Nations Charter and the OSCE governing principles, as well as [other international treaties]. . . . Russia will exercise similar restraint in its conventional force deployments in Europe.

Russia, on the other hand, has broken its international commitments, including basic principles in the NATO-Russia Founding Act, such as,

refraining from the threat or use of force against each other as well as against any other state, its sovereignty, territorial integrity or political independence in any manner inconsistent with the United Nations Charter and with . . . [other international treaties] and the "respect for sovereignty, independence and territorial integrity of all states and their inherent right to choose the means to ensure their own security, the inviolability of borders and peoples' right of self-determination" as enshrined in [several international treaties].

Between March and May 2014, Russia had massed around 40,000 troops on Ukraine's border and threatened to invade Ukraine. As of 11 July 2014, Russia still has around 12,000 troops, tanks and artillery close to the Ukrainian border. Over the past months, Russia has also embarked on an unprecedented schedule of no-notice military exercises involving massive numbers of troops and heavy equipment. Russia should explain what its military plans are before it starts accusing others of posing a threat.

Secondly, all of NATO's deployments have taken place on NATO territory, with the intention to deter threats to NATO territory. Russia, on the other hand, has illegally annexed Crimea, allowed mercenaries and heavy weapons to flow across its border into Ukraine, and refused to condemn the aggressive and illegal actions of armed separatists in Ukraine, as it committed to do in Geneva in April. Recruiting efforts for separatist fighters are also expanding inside Russia.

NATO is showing strict respect of international borders and international commitments. Russia should do the same.

Russian Claims That the So-Called Referendum in Crimea Was Legal

Russian officials claim that the so-called referendum in Crimea on 16 March was legal.

The referendum was illegal according to the Ukrainian constitution, which states that questions "of altering the territory of Ukraine are resolved exclusively by an All-Ukrainian referendum." Crimea, as part of Ukraine, has the status of an autonomous republic, but any issues about its authority have to be resolved by the Ukrainian parliament (Article 134) and its constitution has to be approved by the Ukrainian parliament (Article 135).

The UN General Assembly on 27 March 2014 passed a non-binding resolution declaring the so-called referendum invalid. The European Union also does not recognize the alleged outcome. Additionally, the so-called referendum was organized in a matter of weeks by a self-proclaimed Crimean leadership that was installed by armed Russian military personnel after seizing government buildings. Obviously, any such fake referenda organized by self-appointed authorities who lack all democratic legitimacy are illegal and illegitimate.

It should be noted that Moscow never lodged a single complaint with any international body about the alleged discrimination of Russian-speaking citizens of Ukraine.

Russian Claim NATO's Continuation and Enlargement Threatens Russia

Russian officials say that NATO should have been disbanded at the end of the Cold War, and that the accession of new Allies from Central and Eastern Europe undermines Russia's security.

NATO was not disbanded after the Cold War because its members wanted to retain the insurance policy that had guaranteed security and stability in the transatlantic area and beyond. As the London Declaration of 1990 [by NATO's members] makes clear: "We need to keep standing together, to extend the long peace we have enjoyed these past four decades." Upholding the values that have always guided it, NATO became more than a powerful military Alliance: it became a political forum for dialogue and cooperation.

NATO has fulfilled the terms of Article 10 of the North Atlantic Treaty, which states that Allies "may, by unanimous agreement, invite any other European State in a position to further the principles of this Treaty and to contribute to the security of the North Atlantic area to accede to this Treaty."

On six occasions, between 1952 and 2009, European countries made the choice to apply for membership based on a democratic process and respect for the rule of law. NATO Allies made the unanimous choice to accept them.

NATO and EU enlargement has helped the nations of Central and Eastern Europe to tackle difficult reforms, which were required prior to accession. It has helped their citizens enjoy the benefits of democratic choice, the rule of law, and substantial economic growth. These efforts have moved Europe closer to being whole, free, and at peace than at any other time in history.

Russia also subscribed to this vision in the Founding Act as well as other documents. It committed to "creating in Europe a common space of security and stability, without dividing lines or spheres of influence," and to "respect for sovereignty, independence and territorial integrity of all states and their inherent right to choose the means to ensure their own security."

Contrary to those commitments, Russia now appears to be attempting to recreate a sphere of influence by seizing a part of Ukraine, maintaining thousands of forces on its borders, and demanding, as Russian Foreign Minister Sergei Lavrov has stated, that "Ukraine cannot be part of any bloc."

On the Claim That NATO Wants to "Drag" Ukraine into NATO

NATO does not "drag" countries into the Alliance. NATO respects the right of every country to choose its own security arrangements. In fact, the Washington Treaty [that established NATO in 1949] specifically gives Allies the right to leave. Over the past 65 years, 28 countries have chosen freely to join NATO. Not one has asked to leave. This is not dragging, it's sovereign choice.

NATO's Open Door policy has been, and will always be, based on the free choice of European democracies. When in 2002 under President [Leonid] Kuchma Ukraine decided to pursue NATO membership, the Alliance took steps to help fulfill Ukraine's Euro-Atlantic aspirations. When in 2010 Ukraine decided to pursue a "non-bloc policy," NATO fully respected that choice. Russia's long-time assertion that NATO tried to force Ukraine into its ranks was, and remains, completely false.

Any decision for Ukraine to apply for membership would have to be taken by Ukraine, in line with its democratic rules. When Foreign Minister [Pavlo] Klimkin was in Brussels in July 2014, he made clear that NATO membership is not on the agenda. The government and people of Ukraine have other priorities. We respect their choices, Russia should do the same.

Claim That Russia Had to Act to Stop NATO from Basing Missiles and Ships in Sevastopol

This is total fantasy. NATO had no intention of deploying forces to Sevastopol. This was never discussed and there have never been any plans for that. The only one who talked about this ludicrous claim was [Russia's] President [Vladirmir] Putin. In fact, before the Ukraine crisis, the only NATO forces routinely present on the territory of Eastern European Allies were the NATO jets used in the Baltic States for the air policing mission.

On the contrary, the only country which had ships and troops in Sevastopol [a port in the Crimea] was Russia, under its agreement with Ukraine. And after the illegal takeover of Crimea, Russia stole most of the ships of the Ukrainian navy and installed additional anti-ship and anti-aircraft batteries to expand its military presence in the region.

Russian Claims That NATO Promised Not to Enlarge or Build Infrastructure in Eastern Europe

Russian officials claim that US and German officials promised in 1990 that NATO would not expand into Eastern and Central Europe, build military infrastructure near Russia's borders or permanently deploy troops there. No such pledge was made, and no evidence to back up Russia's claims has ever been produced.

Should such a promise have been made by NATO as such, it would have to have been as a formal, written decision by all NATO Allies. Furthermore, the consideration of enlarging NATO came years after German reunification. This issue was not yet on the agenda when Russia claims these promises were made. The key question Russia should answer is why so many countries, particularly those on its periphery, continue to aspire to join NATO.

Allegations about NATO pledging not to build infrastructure close to Russia are equally inaccurate. As noted above, in the Founding Act, NATO stressed that:

> In the current and foreseeable security environment, the Alliance will carry out its collective defense and other missions by ensuring the necessary interoperability, integration, and capability for reinforcement rather than by additional permanent stationing of substantial combat forces. Accordingly, it will have to rely on adequate infrastructure commensurate with the above tasks.

NATO has indeed supported the upgrading of military infrastructure, such as air bases, in the countries which have joined the Alliance, commensurate with the requirements for reinforcement and exercises. The process was transparent to all, including Russia. However, the only combat forces permanently stationed on the territory of the new members are their own armed forces.

Even before the Ukraine crisis, the only routinely visible sign of Alliance forces on the territory of new members were the NATO jets used in the Baltic States [Estonia, Latvia, and Lithuania] for the air policing mission. These minimal defensive assets cannot be described as substantial combat forces in the meaning of the Founding Act. By contrast, in 2007, Russia unilaterally suspended its compliance with and later on withdrew from the only comprehensive and verifiable arms control regime in Europe, the Treaty on Conventional Armed Forces in Europe.

Since the crisis, NATO has taken steps to increase situational awareness and bolster the defenses of our Eastern members. This, too, is entirely consistent with the Founding Act and is a direct result of Russia's destabilizing military actions.

Finally, the Act also states, "Russia will exercise similar restraint in its conventional force deployments in Europe." Russia's aggression against Ukraine is a flagrant breach of this commitment.

Russian Claims That NATO Missile Defense Is a Threat to Russia

Russia's arguments that NATO missile defense could undermine Russia's strategic deterrent are baseless. NATO's missile defense is neither designed nor directed against Russia. It is designed and located to defend NATO population and territories against threats from outside the Euro-Atlantic area.

Moreover, the Alliance has consistently sought cooperation with Russia on missile defense. At the Lisbon Summit of 2010, NATO Heads of State and Government "decided to develop a missile defence capability to protect all NATO European populations, territory and forces, and invited Russia to cooperate with us."

This was reiterated at the Chicago Summit in May 2012, where leaders underlined that NATO "remains committed to cooperation on missile defense in a spirit of mutual trust and reciprocity," and stated explicitly that NATO missile defense "will not undermine Russia's strategic deterrence capabilities." NATO also proposed a transparency regime including the creation of two NATO–Russia joint missile-defense centers. Russia has declined these offers.

These Summit declarations are more than political promises: they define NATO's policies. Rather than taking NATO up on cooperation, Russia has advanced arguments that ignore laws of physics as well as NATO's expressed policies. Independent Russian military experts have made clear that NATO's missile defense program could not pose any threat to Russia or degrade the effectiveness of its strategic deterrent forces. The Russian government has used missile defence as an excuse for accusations rather than an opportunity for partnership.

Russian Claims That the United States Is Disinterested in Europe and That the Alliance Is Not United

Russian officials claim that the United States is no longer interested in the security of Europe. This is simply false. Every single Ally is interested in Europe's security, and every single Ally is contributing.

Since the crisis began, U.S. soldiers have deployed to the Baltic States—alongside European troops. U.S. ships have sailed in the Baltic, Mediterranean and Black Seas, alongside European and Canadian vessels. U.S. aircraft have policed the skies of Eastern Europe, alongside European and Canadian planes. President Obama's announcement of a European Reassurance Initiative of up to 1 billion dollars to further reinforce NATO's collective defense underscores the United States' unwavering commitment to NATO.

The Alliance is also looking into long-term measures to enhance the security of all member states in view of Russia´s actions. Every single member of NATO is contributing to the Alliance's response to this crisis. There is no stronger proof of the unity of NATO—and the inaccuracy of Russia's claims.

NATO–Russia Relations

Russia claims that NATO has spent years trying to marginalize it internationally.

Since the early 1990s the Alliance has consistently worked to build a cooperative relationship with Russia on areas of mutual interest, and striven towards a strategic partnership.

Before the fall of the Soviet Union and the Warsaw Pact [between the Soviet Union and its East European allies], NATO began reaching out, offering dialogue in place of confrontation, as the London NATO Summit of July 1990 made clear. In the following years, the Alliance promoted dialogue and cooperation by creating [several] new forums, open to the whole of Europe, including Russia.

As a sign of Russia's unique role in Euro-Atlantic security, in 1997 NATO and Russia signed the Founding Act on Mutual Relations, Cooperation and Security, creating the NATO–Russia Permanent Joint Council. In 2002 they upgraded that relationship, creating the NATO–Russia Council (NRC).

Since the foundation of the NRC, NATO and Russia have worked together on issues ranging from counter-narcotics and counter-terrorism to submarine rescue and civil emergency planning. No other partner has been offered a comparable relationship, nor a similar comprehensive institutional framework. Far from marginalising Russia, NATO has treated it as a privileged partner. By contrast, Russia has referred to NATO as a threat in its strategic documents.

Russian Criticism of the Legitimacy of NATO Military Actions—Kosovo

The NATO operation for Kosovo followed over a year of intense efforts by the UN and the Contact Group, of which Russia was a member, to bring about a peaceful solution. The UN Security Council on several occasions branded the ethnic cleansing in Kosovo and the mounting number of refugees driven from their homes as a threat to international peace and security. NATO's Operation Allied Force was launched to prevent the large-scale and sustained violations of human rights and the killing of civilians.

Following the air campaign, the subsequent NATO-led peacekeeping operation, KFOR, which initially included Russia, has been under UN mandate, with the aim of providing a safe and secure environment in Kosovo. This led to nearly ten years of diplomacy, under UN authority, to find a political solution and to settle Kosovo's final status.

The Kosovo operation was conducted following exhaustive discussion involving the whole international community dealing with a long-running crisis. In Crimea, with no evidence of a crisis and no attempt to negotiate any form of solution, Russia bypassed the whole international community, including the UN, and simply occupied a part of another country's territory.

Russian Claims That the Annexation of Crimea Was Justified by the Opinion of the International Court of Justice on the Independence of Kosovo

Russian leaders claim that the precedent for the so-called declaration of independence of Crimea was the advisory opinion of the International Court of Justice on the independence of Kosovo.

However, the court stated clearly that their opinion was not a precedent. The court said they had been given a "narrow and specific" question about Kosovo's independence, which would not cover the broader legal consequences of that decision.

The court highlighted circumstances in which claims for independence would be illegal. This would include if "they were, or would have been, connected with the unlawful use of force." An example of "an unlawful use of force" would be an invasion and occupation by a neighboring country—which is exactly what Russia has done.

Furthermore, the process leading to Kosovo's declaration of independence spanned years and included an extensive process led by the United Nations. Russian claims ignore all of these facts.

THE NORTH ATLANTIC TREATY ORGANIZATION is a trans-Atlantic defense treaty that was established in 1949 and currently has 28 member countries, including the United States and Canada.

EXPLORING THE ISSUE

Is Russia Being Demonized for Pursuing Its National Interests?

Critical Thinking and Reflection

1. It is important to understand how an opponent sees a situation to avoid mistaking an opponent's motives. Imagine you are Russian. Your country went through two horrific world wars during the twentieth century and faced enmity from the United States and most of Western Europe for some seventy years because of its communist government. Now those same countries have extended their two most powerful international organizations, NATO and the EU, to your borders to include most of your country's cold war allies in Eastern Europe and even three former Soviet republics. Those three are small, but NATO and the EU have also begun to discuss relations with the far more important FSR, Ukraine. Is it possible to accept at least some of the view of Putin that Russia is acting out of self-defense when it reacts against the pro-Western direction Ukraine has taken in recent years? If so, how, if any, does that view change your view of what U.S., NATO, and EU policy should be?
2. Should it make any difference in U.S. foreign policy toward Russia whether or not the country is authoritarian or democratic?
3. Some analysts argue that it is naïve to not recognize Russia as hostile and that failing to do so can lead into danger in such ways as making arms agreements that Russia will not keep. Other analysts argue that treating Russia as hostile will become a self-fulfilling prophecy by prompting Russia to react with hostility to U.S. policy proposals. Which position do you find more plausible?

Is There Common Ground?

The debate over what motives drive Russia and how to deal with it are not matters of idle speculation. There are two very real policy considerations. The first involves the fact that the direction Russia takes in the future is likely to have important consequences for the world. During the decade or so after the collapse of the Soviet Union in 1991, Russia was in great economic and military disarray. President Boris Yeltsin and President Vladimir Putin both criticized such U.S.-favored actions as the expansion of NATO, but Moscow's weakness constrained it from trying to block U.S. preferences.

More recently, Russia's economic fortunes have improved. The global demand for energy and the increasing revenue that Russia gets from its oil and natural gas exports have given a particular boost to Russia's economy. It is yet unclear how much Russia's military has recovered, but there are some signs that its morale, training, and operational capability are on the upswing.

All this has given Russia greater confidence to assert itself when its policy choices cause conflict with the United States and Western Europe.

As recently as 2012, the Obama administration voiced support for "resetting" relations with Russia; that is, looking for ways to cooperate with Russia. In June 2012,

Deputy National Security Advisor for Strategic Communications Ben Rhodes contended, "The reset with Russia is based on the belief that we can cooperate with them on areas of common interest, understanding that we still have some differences. And I think we continue to see very positive results from that reset policy." Others disagreed. The campaign site of Republican presidential candidate Mitt Romney labeled Obama's reset policy a failure and pledged, "Upon taking office [as president], Mitt Romney will reset the reset. He will implement a strategy that will seek to discourage aggressive or expansionist behavior on the part of Russia and encourage democratic political and economic reform." More recently, the idea of a reset has vanished from the debate in the United States, and American public opinion has come to see Russia as hostile. As recently as 2006, the Gallup Poll found 73 percent of Americans viewing Russia as "friendly or an ally," and only 20 percent depicting Russia as "unfriendly or an enemy." In 2014 that view has changed dramatically, with only 26 percent of Americans viewing Russia as "friendly or an ally," and only 58 percent depicting Russia as "unfriendly or an enemy." Views in Russia of the United States are similar. Only 23 percent of Russians view the United States favorable in 2014 according to the Pew Research Center. That is down from 61 percent in 2002.

Does this mean that relations with Russia will continue to plummet and that conflict, even war, is in the future? It does not. Both sides must look for ways to allow the countries on Russia's borders to cooperate with the EU and even NATO if they wish while, at the same time, not creating agreements or other ties that alarm Russia in the same way that close ties between Russia or China and countries in the Western Hemisphere would alarm and be opposed by the United States. It is also the case that there are strong reasons for both sides to cooperate. Western Europe relies heavily on energy supplies from Russia, and the sale of natural gas and other energy supplies are a key part of trade income that Russia needs to support its economy. Moreover, costs of any military conflict between Russia and the West would be heavy and could even pale the devastation of World War II. You would think that would be enough to ensure peace, but it is not. Both sides need to moderate their goals and reactions.

Create Central

www.mhhe.com/createcentral

Additional Resources

Galeotti, Mark, and Bowen, Andrew S. (April 21, 2014). "Putin's Empire of the Mind." *Foreign Policy* online at: http://www.foreignpolicy.com/articles/2014/04/21/putin_s_empire_of_the_mind_russia_geopolitics

Galeotti and Bowen offer a negative view of Putin as intent on restoring the old czarist/Soviet empire.

Gvosdev, Nikolas K., and Marsh, Christopher (2013). *Russian Foreign Policy: Interests, Vectors, and Sectors* (CQ Press)

This study gives a recent overview of Russian foreign policy.

Leichtova, Magda (2014). *Misunderstanding Russia: Russian Foreign Policy and the West* (Ashgate Publishing)

This book takes a less negative view of Russia than Galeotti and Bowen above.

Internet References . . .

Moscow Times

For Russia's slant on the news, read one of the country's online newspapers in English, such as the *Moscow Times*.

www.themoscowtimes.com/

Foreign Policy Concept of the Russian Federation (2013*)*

An explanation of Russia foreign policy published by the Russian Foreign Ministry

http://www.ieee.es/en/Galerias/fichero/docs_marco/2013/DIEEEM06-2013_Rusia_ConceptoPoliticaExterior_FRuizGlez_ENGLISH.pdf

Country Studies: Russia

The U.S. Department of State's website on Russia

http://www.state.gov/p/eur/ci/rs/

Web Portal of Ukrainian Government

http://www.kmu.gov.ua/control/en

Selected, Edited, and with Issue Framing Material by:
John T. Rourke, *University of Connecticut, Storrs*

ISSUE

Is China Being the Aggressor in the Issue of the Disputed Island in the Pacific Ocean?

YES: Chuck Hagel, from "The United States' Contribution to Regional Stability," An Address at the Shangri-La Dialogue, the 13th International Institute for Strategic Studies Asia Security Conference, Singapore (2014)

NO: Wang Guanzhong, from "Major Power Perspectives on Peace and Security in the Asia-Pacific," Remarks during the Fourth Plenary Session of the Shangri-La Dialogue, International Institute for Strategic Studies, Singapore (2014)

Learning Outcomes

After reading this issue, you will be able to:

- Elaborate on the background to the dispute between China and its neighbors over various islands and ocean regions in the far eastern Pacific Ocean.
- Explain the position of the United States on the disputes.
- Comment on how the disputes are affecting military spending and the foreign policy of the countries in the region.

ISSUE SUMMARY

YES: Chuck Hagel, the U.S. secretary of defense, tells delegates gathered in Singapore in 2014 at the Shangri-La Dialogue that China has undertaken destabilizing, unilateral actions in aggressively asserting its claims to significant areas in the South China Sea and the East China Sea and to many of the islands in them, the resources in and under those areas, and even to maritime and aviation transit through and over them. The Shangri-La Dialogue is a series of conferences that has been held annually since 2002 and is sponsored by the International Institute for Strategic Studies (IISS), a private research and advocacy organization headquartered in London. The conference brings together representatives of the defense ministries of countries located in or bordering on the Pacific Ocean and also several European countries including France, Germany, and Great Britain to discuss security in the Pacific region.

NO: Wang Guanzhong, a lieutenant general and deputy chief of staff of the People's Liberation Army, China, responds to Secretary of Defense Hagel's speech and a similar, if less blunt, speech by Prime Minister Shinzo Abe of Japan by condemning them for making unwarranted accusations against China and thereby provoking and challenging China. General Wang asserts that China is ready to cooperate with other countries in the greater Pacific region to promote regional and global peace and development.

China has a history as one of the oldest and at times most powerful countries (and empires) in the world. During the Yuan dynasty (1271–1368) and most of the Ming dynasty (1368–1644), China was also arguably the world's most powerful empire, dominating most of Asia.

However, China's power compared with Europe's began to ebb, with the Industrial Revolution beginning in Europe in the mid-1700s playing a major role. During the 1800s the European powers, joined by the United States in the last years of that century, increasingly dominated China. The Chinese consider these years a period of humiliation. Their resentment was epitomized by the Boxer Rebellion (1898–1900), during which Chinese militias tried to drive out foreigners from their country but were ultimately defeated by an international force that included troops from Austria-Hungary, France, Germany, Great Britain, Italy, Japan, Russia, and the United States.

China's road back began in 1911 when Nationalist forces under Sun Yatsen overthrew the last emperor. Internal struggles and the invasion by Japan (1931–1945) blocked much advance in China's economic and political power until the Communists under Mao Zedong defeated the Nationalists under Chiang Kai-shek, who fled and set up the remnants of the Nationalist government on Formosa (Taiwan) as the Republic of China. Gradually, Communist China (the People's Republic of China, PRC) built up its strength. Military power came first. China's military was limited by obsolete weapons, but it was the world's largest military force, with as many as 4.2 million troops. China also sought to acquire nuclear weapons and delivery capability, and succeeded in that quest by the mid-1960s. Basic changes in China's global status began in the 1970s. In 1971, the United Nations changed the rightful owner of China's seat, including its position as a permanent member of the Security Council, from the Nationalist government on Taiwan to the PRC. The following year, the United States relaxed its hostility, and President Richard Nixon visited China. In 1979, President Jimmy Carter shifted U.S. diplomatic recognition of the "legitimate" government of China from the Taiwan government to the PRC government in Beijing.

Additionally, the two great leaders of the Communist Revolution and government, Premier Zhou Enlai and Communist Party Chairman Mao Zedong, both died in 1976. This opened the way for a less ideological approach to China's domestic and foreign policy. Since then, China has changed rapidly. It retains a communist government, but it has adopted many of the trappings of a capitalist economy. Where once China rejected global trade and other international economic organizations, it has now embraced them.

How powerful China has become economically depends on how you count its gross domestic product (GDP). On a per capita basis in 2013, it was only $6,900 compared to $53,000 for the United States. But on an overall basis, China's 2013 GDP was $9.2 trillion, second in the world only to that of the United States ($16.8 trillion).

China's rapidly growing and modernizing economy has allowed it to continue upgrading its military technology. China is also second among countries in military spending. The country's 2013 official defense budget was $188 billion dollars, but there is little doubt that actual spending is higher than that, with some estimates putting it nearer $225 billion. Still the amount, whatever it is, falls far short of the 2013 U.S. defense budget ($640 billion). It is also noteworthy that China's military spending has been rising rapidly, increasing 12.2 percent between 2013 and 2014 alone. This is part of a worrisome "arms race" in the Asian region. While global defense spending controlled for inflation declined in 2013, it increased 3.6 percent in the region.

China's growing economic and military power has provided it with the foundation to continue to increase its position as a global and, even more, as a regional power. China claims that it owns or has authority over parts of the eastern Pacific Ocean that are, in some part, claimed by Brunei, Japan, Indonesia, Malaysia, Philippines, Taiwan, and Vietnam. China's assertion of the right to control fishing rights and activity in and even transit through or over other areas has also created friction with other countries, including the United States, that consider some of the areas to be international waters. The year 2014 saw increased tension in the area as China and other claimant countries asserted their rights by using their naval forces to assert control, and the United States used its naval forces to assert its right to unimpeded transit.

YES

Chuck Hagel

The United States' Contribution to Regional Stability

Last year, I participated in this dialogue during my first visit to the Asia-Pacific as [U.S.] Secretary of Defense. As a United States Senator I have been here many times. I spoke about the United States of America's firm commitment—firm commitment to this region's security and economic prosperity, and to supporting its extraordinary progress through our strategic rebalance.

Today, I return on my fifth trip to the region as Secretary of Defense in about a year, again reaffirming that America's commitment to the Asia-Pacific is enduring.

In his remarks at West Point earlier this week, President [Barack] Obama laid out the next phase of America's foreign policy—particularly as we come out of 13 years of war in Iraq and Afghanistan. He made clear we will balance our diplomacy, our development assistance, and military capabilities, and that we will strengthen our global partnerships and alliances.

That is how America is implementing its strategy of rebalancing to the Asia-Pacific. The rebalance is not a goal, not a promise, or a vision—it's a reality. Over the last year, President Obama launched comprehensive partnerships with Vietnam and Malaysia, held a summit with Chinese President Xi [Jinping], and last month visited three of our five regional treaty allies—Japan, South Korea, and the Philippines—as well as Malaysia. In the Philippines, he and President Aquino announced a new Enhanced Defense Cooperation Agreement on the rotational presence of U.S. forces—the most significant milestone for our alliance in over a decade.

Under President Obama's leadership, the administration is also making progress in negotiating the Trans-Pacific Partnership trade agreement. Our State Department is increasing foreign assistance funding to the Asia-Pacific region and expanding assistance for maritime capacity-building in Southeast Asia.

Diplomatic, economic, and development initiatives are central to the rebalance, and to our commitment to help build and ensure a stable and prosperous region. But prosperity is inseparable from security, and the Department of Defense will continue to play a critical role in the rebalance—even as we navigate a challenging fiscal landscape at home.

A central premise of America's strategy in the Asia-Pacific is our recognition that, in the 21st century, no region holds more potential for growth, development, and prosperity than this one.

But even while advances in human rights, freedom, democracy, technology, and education are all yielding better lives and futures for all people; and even as more nations are stepping forward to contribute to regional security, the Asia-Pacific is also confronting serious threats.

We see ongoing territorial and maritime disputes in the South and East China Seas; North Korea's provocative behavior and its nuclear weapons and missile programs; the long-term challenge of climate change and natural disasters; and the destructive and destabilizing power of cyber attacks.

Continued progress throughout the Asia-Pacific is achievable, but hardly inevitable. The security and prosperity we have enjoyed for decades cannot be assured unless all nations—all our nations—have the wisdom, the vision, and will to work together to address these challenges.

As President Obama said earlier this week, "America must always lead on the world stage. If we don't, no one else will." He went on to say that, the "question is not whether America will lead, but how we will lead . . . to help ensure peace and prosperity around the globe." Today, I want to highlight four broad security priorities that the United States, as a Pacific power, is advancing in partnership with friends and allies throughout the Asia-Pacific:

- First, encouraging the peaceful resolution of disputes; upholding principles including the freedom of navigation; and standing firm against coercion, intimidation, and aggression;
- Second, building a cooperative regional architecture based on international rules and norms;

- Third, enhancing the capabilities of our allies and partners to provide security for themselves and the region; and,
- Fourth, strengthening our own regional defense capabilities.

One of the most critical tests facing the region is whether nations will choose to resolve disputes through diplomacy and well-established international rules and norms . . . or through intimidation and coercion. Nowhere is this more evident than in the South China Sea, the beating heart of the Asia-Pacific and a crossroads for the global economy.

China has called the South China Sea "a sea of peace, friendship, and cooperation." And that's what it should be.

But in recent months, China has undertaken destabilizing, unilateral actions asserting its claims in the South China Sea. It has restricted access to Scarborough Reef, put pressure on the long-standing Philippine presence at the Second Thomas Shoal, begun land reclamation activities at multiple locations, and moved an oil rig into disputed waters near the Paracel Islands.

The United States has been clear and consistent. We take no position on competing territorial claims. But we firmly oppose any nation's use of intimidation, coercion, or the threat of force to assert those claims.

We also oppose any effort—by any nation—to restrict overflight or freedom of navigation—whether from military or civilian vessels, from countries big or small. The United States will not look the other way when fundamental principles of the international order are being challenged. We will uphold those principles. We made clear last November that the U.S. military would not abide by China's unilateral declaration of an Air Defense Identification Zone in the East China Sea, including over the Japanese-administered Senkaku Islands. [China calls them the Diaoyu Islands.] And as President Obama clearly stated in Japan last month, the Senkaku Islands are under the mutual defense treaty with Japan.

All nations of the region, including China, have a choice: to unite, and recommit to a stable regional order, or to walk away from that commitment and risk the peace and security that have benefitted millions of people throughout the Asia-Pacific, and billions around the world. The United States will support efforts by any nation to lower tensions and peacefully resolve disputes in accordance with international law.

We all know that cooperation is possible. Last month, 21 nations signed the Code for Unplanned Encounters at Sea—an important naval safety protocol. ASEAN [the 10-member Association of South East Asian Nations] and

China are negotiating a Code of Conduct for the South China Sea—and the United States encourages its early conclusion. Nations of the region have also agreed to joint energy exploration; this month, the Philippines and Indonesia resolved a longstanding maritime boundary dispute; and this week, Taiwan and the Philippines agreed to sign a new fisheries agreement. China, too, has agreed to third-party dispute resolution in the World Trade Organization; peacefully resolved a maritime boundary dispute with Vietnam in 2000; and signed ASEAN's Treaty of Amity and Cooperation.

For all our nations, the choices are clear, and the stakes are high. These stakes are not just about the sovereignty of rocky shoals and island reefs, or even the natural resources that surround them and lie beneath them. They are about sustaining the Asia-Pacific's rules-based order, which has enabled the people of this region to strengthen their security, allowing for progress and prosperity. That is the order the United States—working with our partners and allies—that is the order that has helped underwrite since the end of World War II. And it is the order we will continue to support—around the world, and here in the Asia-Pacific.

This rules-based order requires a strong, cooperative regional security architecture. Over the last year, the United States has worked with Asia-Pacific nations to strengthen regional institutions like ASEAN and the ADMM+ [meetings of the ASEAN defense minister plus counterparts from eight other countries, including the United States], which I attended last year in Brunei. This regional architecture is helping to develop shared solutions to shared challenges, building strong and enduring ASEAN security community, and ensuring that collective, multilateral operations are the norm, rather than the exception.

A common picture of the region's maritime space could help deter provocative conduct, and reduce the risk of accidents and miscalculation. So I am asking Admiral Sam Locklear, who leads the United States Pacific Command, to host his regional counterparts to discuss concrete ways to establish greater maritime security awareness and coordination.

The United States is also reaching out to China. We're reaching out to China because we seek to expand prosperity and security for all nations of this region. As I underscored in Beijing last month during my visit to China, the United States will continue to advance President Obama and President Xi's shared commitment to develop a new model of relations—a model that builds cooperation, manages competition, and avoids rivalry. To help develop this model, we are increasing our military-to-military engagement with China through our joint exercises, exchanges, and other confidence-building measures that can help improve communication and build understanding between our forces. Chairman of the [U.S.] Joint Chiefs General [Martin] Dempsey and I have led this effort, and we will continue to focus on building this new military-to-military model. And I am glad General Dempsey is here to help us today accomplish more progress in this area.

The U.S.-China military-to-military dialogue has a long way to go. But I think we've been encouraged by the progress we've made, and continue to make. Our dialogue is becoming more direct, more constructive . . . getting at the real issues and delivering more results. As we expand this dialogue, the United States also supports a sustained and substantive exchange with China on cyber issues. Although China has announced a suspension of the U.S.-China Cyber Working Group, we will continue to raise cyber issues with our Chinese counterparts, because dialogue is essential for reducing the risk of miscalculation and escalation in cyberspace.

The Asia-Pacific's shifting security landscape makes America's partnerships and alliances indispensable as anchors for regional stability. As we work to build a cooperative regional architecture, we are also modernizing our alliances, helping allies and partners develop new and advanced capabilities, and encouraging them to work more closely together.

In Southeast Asia, that means continuing to help nations build their humanitarian and disaster relief capabilities, and upgrade their militaries. One important example is our first-ever sale of Apache helicopters to Indonesia, which I announced during my visit to Jakarta last year. This sale will help the Indonesian Army defend its borders, conduct counter-piracy operations, and control the free flow of shipping through the Straits of Malacca. We are also providing robust assistance to the Philippines' armed forces, to strengthen their maritime and aviation capabilities. In Northeast Asia, our capacity-building efforts include strengthening Allies' capabilities with sophisticated aircraft and ballistic missile defense—especially to deter and defend against provocation by Pyongyang.

We signed that agreement. We are also making significant progress in building a robust regional missile defense system. Last month in Tokyo, I announced that the United States will deploy two additional ballistic missile defense ships to Japan—a step that builds on the construction of a second missile defense radar site in Japan, and the expansion of America's ground-based interceptors in the continental United States, which I reviewed this week in Alaska during my trip to Singapore.

Modernizing our alliances also means strengthening the ties between America's allies, enhancing their joint capabilities—such as missile defense—and encouraging them to become security providers themselves. Yesterday, I held a trilateral meeting with my counterparts from Australia and Japan, and today I will host another trilateral meeting with my counterparts from Korea and Japan.

The enhanced cooperation America is pursuing with these close allies comes at a time when each of them is choosing to expand their roles in providing security around the Asia-Pacific region, including in Southeast Asia. Seven decades after World War II, the United States welcomes this development. We support South Korea's more active participation in maritime security, peacekeeping, and stabilization operations. We also support Japan's new efforts—as Prime Minister Abe described very well last night—to reorient its Collective Self Defense posture toward actively helping build a peaceful and resilient regional order.

To complement these efforts, the United States and Japan have begun revising our defense guidelines for the first time in more than two decades. We will ensure that our alliance evolves to reflect the shifting security environment, and the growing capabilities of Japan's Self-Defense Forces.

The United States also remains committed to building the capacity of allies and partners in the region through as many as 130 exercises and engagements, and approximately 700 port visits annually. And across the Asia-Pacific region, as part of the rebalance, the United States is planning to increase Foreign Military financing by 35%, and military education and training by 40% by 2016.

Next month, the United States will host its annual Rim of the Pacific exercise, the world's largest maritime exercise that will feature for the first time a port visit by a New Zealand naval ship to Pearl Harbor in more than 30 years, and it will include Chinese ships for the first time. Beyond capacity-building efforts, a stable and peaceful regional order depends on a strong American military presence across the Asia-Pacific region . . . a presence that enables us to partner with our friends and allies, and help deter aggression. We are no strangers to this part of the world. America has been a Pacific power for many years. Our interests lie in these partnerships and this region.

Today, America has more peacetime military engagement in the Asia-Pacific than ever before. I want to repeat: today, America has more peacetime military engagement in the Asia-Pacific than ever before. And America's strong military presence—and our role in underwriting the region's security—will endure. Our friends and allies can judge us on nearly seven decades of commitment and history of commitment. That history makes clear, America keeps its word. America's treaty alliances remain the backbone of our presence in the Asia-Pacific, and our friends and allies have seen our significant steps in recent years to enhance our posture in Northeast Asia, to expand our partnerships in Southeast Asia, and to ensure our forces can operate effectively regardless of other nations' capabilities.

In the coming years, the United States will increase its advanced capabilities that are forward-stationed and forward-deployed in the entire region, particularly as we draw down our forces in Afghanistan. And we will ensure that we sustain our freedom of action in the face of disruptive new military technologies.

Finally, to ensure that the rebalance is fully implemented, both President Obama and I remain committed to ensuring that any reductions in U.S. defense spending do not come—do not come—at the expense of America's commitments in the Asia-Pacific.

Here, and around the world, a peaceful, prosperous, and durable order will not sustain itself. The nations of the Asia-Pacific must come together to accomplish this.

We must support the peaceful resolution of disputes . . . and oppose intimidation and coercion no matter where they are.

We must build a cooperative regional security architecture that builds trust and confidence. And we must continue to develop, share, and maintain advanced military capabilities that can adapt to rapidly growing challenges.

From Europe to Asia, America has led this effort for nearly seven decades, and we are committed to maintaining our leadership in the 21st century. Across this region, and across the globe, the United States has been—and always will be—committed to a peaceful and prosperous international order that rests not merely on America's own might, but on our enduring unity and partnership with other nations.

CHUCK HAGEL is the U.S. secretary of defense and a former U.S. Democratic senator from Nebraska (1997–2009). He received a BA degree in history from the University of Nebraska at Omaha.

Wang Guanzhong

 **NO**

Major Power Perspectives on Peace and Security in the Asia-Pacific

Today, I'd like to take the opportunity to share with you the values, policies and practices of China for maintaining peace and security in the Asia-Pacific region.

Not long ago, the fourth Summit of the Conference on Interaction and Confidence-Building Measures in Asia was held in Shanghai, China. At the Summit, Chinese President Xi Jinping put forth the security concept for Asia featuring common, comprehensive, cooperative and sustainable security. This concept, which is a profound summary of the Asian historical experiences as well as the cherished aspiration for Asia's future, has been widely acclaimed by the Asian countries. Asia today is in a critical period of development. Asia is increasingly becoming a community of common interest, destiny and responsibility. The stability of Asia is a blessing for world peace and the rejuvenation of Asia is a boon for world development.

The security of China is closely linked to that of Asia. China advocates and implements the security concept for Asia in real earnest, and stands ready to work with other countries to pursue Asian security that is established, shared by, and win-win to all. China is a constructive, proactive and positive force for Asia's peace and security.

China pursues the path of peaceful development. Peaceful development is a strategic choice as well as a long-term and abiding strategy made by China, based on its historical and cultural traditions, historical experiences, lessons learned in the rise of major powers in the past, the reality of our time and the fundamental interests of China. China sticks to open development, cooperative development and win-win development. China strives for self-development through maintaining a peaceful international environment, and in turn contributes to regional and world peace with its own development. China is committed to building a harmonious Asia as well as an amicable world of lasting peace and common prosperity. China will never contend for or seek hegemony and foreign expansion. China adheres to peaceful development, which is its major contribution to security in Asia. The

tremendous achievements of China's peaceful development constitute a positive factor of critical importance to the security of Asia.

China upholds the banner of fairness and justice. China believes that all countries, regardless of size, wealth or strength, should have the equal rights to independently choose their own social systems and development paths. We need to learn from each other to offset our own shortcomings and oppose interference in other countries' internal affairs. All countries should enjoy equal participation in regional security affairs. We need to strengthen coordination on the basis of mutual respect, and oppose attempt by any country to dominate regional security affairs. All countries should respect and accommodate the legitimate security concerns of others and enjoy common security through mutual accommodation. We oppose the practices of flexing up military alliances against a third party, resorting to the threat or use of force, or seeking so-called absolute security of one's own at the cost of the security of others.

China advocates dialogue and cooperation. China upholds that all countries should enhance strategic mutual trust, reduce misgivings and coexist in harmony through dialogue and communication. We should continuously strengthen and expand areas of cooperation, take innovative approaches, and seek peace and security through cooperation. All countries should respect each other's sovereignty, independence and territorial integrity, and resolve disputes peacefully through negotiations.

China stands for coordinated progress of security and development. In China's perspective, development lays the foundation for security, which in turn provides the conditions for development. Development is the most important security and the master key to resolving Asia's security issues. China pursues a neighborhood diplomacy that aims at bringing harmony, security and prosperity to its neighbors. China practices the principles of amity, sincerity, mutual benefit and inclusiveness. We work to promote the sound interaction between regional economic cooperation and security cooperation, and to maintain both traditional and non-traditional security in a coordinated way. In 2013, China contributed nearly 30% of the world's economic growth and over 50% of the growth in Asia. China will continue to promote sustainable security through sustainable development, and work together with other countries for lasting peace and prosperity in the region.

China always pursues a defense policy that is defensive in nature. The PLA [People's Liberation Army] is endeavoring to contribute to maintaining regional security.

We actively conduct friendly military exchanges and cooperation with countries in the Asia-Pacific. China's military-to-military cooperation with other Asia-Pacific countries is showing unprecedented dynamism. We have established defense consultation and dialogue mechanisms with 13 neighboring countries. In recent years, we have held over 50 joint exercises and drills with other Asia-Pacific countries. A sound momentum has been witnessed in high-level visits, professional exchanges and personnel training. Overall, we are enjoying military cooperation in the Asia-Pacific that covers all dimensions, broad areas and multiple levels. We continue to add to the security contents of the comprehensive strategic coordinative partnership with Russia, build towards a new model of military-to-military relationship with the United States, and enhance friendly military security cooperation with India and other major Asia-Pacific countries. The robust military cooperation among major powers plays an important role in maintaining regional security.

We extensively participate in regional multilateral defense and security cooperation. We are actively engaged in security cooperation within Shanghai Cooperation Organization, the Conference on Interaction and Confidence-Building Measures in Asia, ADMM-plus, the ASEAN Regional Forum and China-ASEAN framework. We make joint efforts to fight terrorism, extremism and separatism, conduct disaster relief operations, safeguard the security of land and sea routes and promote a security architecture in line with the common interests of regional countries. Not long ago, we successfully hosted the biennial meeting of the Western Pacific Naval Symposium. Participants of the meeting agreed on the revised version of the Code for Unalerted Encounters at Sea, contributing to the prevention of accidents at sea in the region.

We are committed to properly handling disputes over territory, sovereignty and maritime rights and interests. China has settled land border demarcation with 12 out of 14 of its neighbors and completed the delimitation of Beibu Gulf with Vietnam. The PLA has set up 64 border meeting stations. Over 2,000 meetings were held between Chinese and neighboring border troops in these stations in 2013. We are actively engaged in maintaining maritime security and stability in the neighboring area. In 2002, China and ASEAN countries signed the Declaration on the Conduct [DOC] of Parties in the South China Sea, and jointly set forth the principle that all disputes over territory and jurisdiction should be resolved peacefully through friendly consultation and negotiations between sovereign states directly involved in the disputes. The PLA actively supports the implementation of the DOC and pushes forward the consultation on the Code of Conduct on the South China Sea, with a view to maintaining security and stability in the South China Sea. While firmly safeguarding its sovereignty and legitimate interests, China has demonstrated utmost sincerity and patience in its commitment to settling disputes peacefully through consultations and negotiations with parties involved. China has never threatened to use force, and has never taken provocative actions. We will never accept provocation by others under the pretext of "Proactive Contribution to Peace" that stirs up tension for their selfish interests.

I would like to depart a bit from my prepared script and share some of my perspectives on the speeches of [Japan's] Prime Minister [Shinzo] Abe and [U.S.] Secretary [of Defense Chuck] Hagel [earlier this week] during the Dialogue. Initially I only planned to deliberate on China's policies and make proposals in my speech, not to debate or argue with others. But, unfortunately, after hearing their speeches, I have to offer some comments.

After Mr. Abe's speech [echoing the tone of Hagel's speech in the yes article], a foreign friend attending that session advised me to "be patient." After Mr. Hagel's speech, he again advised me to "be patient." He said that, as for where a country is heading for, what matters is action rather than rhetoric. I believe that was well and correctly said. But today, I have to apologize to this friend of mine that I still have to say a few words in spite of his two-time advice. I will only make a few remarks, in contrast with the lengthy remarks by Mr. Abe and Mr. Hagel to condemn China. I have two considerations. First, as a Chinese proverb goes, it is not polite not to reciprocate. Second, this dialogue is meant to be a forum for everyone to discuss and speak out. And truth can emerge from discussions and debate. Since Mr. Abe and Mr. Hagel have voiced their views about China, I would like to comment on their views, as a way of discussion.

I think the Chinese delegation, the other Chinese as well as many foreign friends attending that session would share my thoughts that the remarks of Mr. Abe and Secretary Hagel staged provocations to China. A foreign friend of mine told me that it was unimaginable for Japanese Prime Minister Shinzo Abe and US Secretary of Defense Mr. Chuck Hagel to make such unwarranted accusations against China. He was right. Those remarks were totally beyond my expectation.

I feel that the speeches of Mr. Abe and Mr. Hagel have been pre-coordinated. They supported and encouraged each other in provoking and challenging China, taking advantage of being the first to speak at the Dialogue [Abe delivered the keynote address, opening the conference]. The focus of Mr. Abe's speech was on China, although he did not name China openly. No matter whether he named China or not and how he tried to whitewash his speech, I believe the entire audience understood that he was targeting China. Mr. Hagel also focused his remarks on China and the entire audience could feel it. Mr. Abe, overtly or covertly, explicitly or implicitly and directly or indirectly condemned China. Mr. Hagel was more frank and straightforward when he made unwanted accusations against China. As for their different approaches and attitude, I would say I prefer those of Mr. Hagel. If you have something to say, say it directly. As an invited government leader, Mr. Abe is supposed to promote peace and security of the Asia-Pacific region with his constructive ideas in line with the principles of the Shangri-La Dialogue. Instead, in violation of those principles, he was trying to stir up disputes and trouble. I do not think this is acceptable or in agreement with the spirit of the Dialogue. Mr. Hagel was more outspoken than I expected. And I personally believe that his speech is a speech with tastes of hegemony, a speech with expressions of coercion and intimidation, a speech with flaring rhetoric that ushers destabilizing factors into the Asia-Pacific to stir up trouble, and a speech with unconstructive attitude. Therefore, one can judge from the two speeches, as well as Mr. Abe and Mr. Hagel's deeds: who is really stirring up trouble and tension in the region and who is initiating disputes and spat? China has never initiated disputes over territorial sovereignty and the delimitation of maritime boundary. China only takes countermeasures against others' provocation. Moreover, China has never initiated provocations on any bilateral or multilateral occasions or at the Shangri-La Dialogue. Who has initiated the ongoing debate? This is well-known to all. Second, from the speeches of Mr. Abe and Mr. Hagel, we know who is really assertive. Assertiveness has come from the joint actions of the United States and Japan, not China.

Such additional comments are simply my passive, reactive and minimum response. Now I will come back to my prepared speech.

Ladies and Gentlemen,

The PLA is ready to work with other militaries to make further contribution to regional and global peace and development. To that end, I would like to make the following proposals:

1. To Promote Mutual Strategic Trust by Deepening Dialogue and Exchanges. We will continue to work with regional countries to carry out in-depth bilateral and multilateral security dialogues and exchanges, and we welcome senior defense officials and scholars of regional countries to join us at the Xiangshan Forum in Beijing in October. China will continue to step up dialogue, communication and coordination with ASEAN countries in the defense and security areas and support the development of the ASEAN Community. Before long, China's State Councilor and Defense Minister General Chang Wanquan invited defense ministers of ASEAN countries to China in 2015 for a Special China-ASEAN Defense Ministers' Meeting. We expect the meeting to achieve significant results.

2. To Support Common Development by Strengthening Security Cooperation. China has proposed to work with regional countries to build a Silk Road Economic Belt and a 21st Century Maritime Silk Road. The two major cooperation initiatives offer new opportunities for China and regional countries to achieve common development. Common development cannot be made without a secure environment. The building of the Economic Belt and the Maritime Silk Road has to be driven by two wheels—development and security. The PLA is ready to work with regional countries to strengthen practical cooperation in counter-terrorism, disaster relief, protection of sea lines of communication and other fields, thus ensuring common prosperity of countries along the Economic Belt and the Maritime Silk Road.

3. To Jointly Tackle Challenges by Promoting Disaster Relief Cooperation. The Asia-Pacific is prone to various disasters. This year, in order to improve regional capacity building, we have arranged five bilateral and multilateral joint disaster relief exercises and drills of all services with regional countries.

4. To Maintain Maritime Security by Highlighting Maritime Cooperation. The ocean serves the common interests of all Asia-Pacific countries. It is therefore the shared responsibility of all to strengthen maritime cooperation and maintain maritime security. China will continue to deepen multi-faceted maritime security cooperation with its regional partners through joint exercises, ship visits and maritime liaison mechanisms.

5. To Effectively Manage Differences by Establishing Security Mechanisms. In order to maintain regional security and stability, it is crucial to properly manage differences, ensure timely communication, and dispel misperceptions and miscalculations.

Ladies and Gentlemen,

Peace does not come easily, and security should be cherished above all. Next year marks the 70th Anniversary of the victory of the world's anti-Fascist war. China will work with all other countries to safeguard the fruits of victory of the Second World War as well as the post-war international order. We will never allow the ruthless Fascist and militarist aggressions to stage a comeback. Major countries shoulder major responsibilities for maintaining security and stability of the Asia-Pacific, while medium and small countries can also play a constructive role. As a responsible major country, China is ready to join hands with all other Asia-Pacific countries to achieve mutual benefit and win-win results. Let us work together to create a better future for the Asia-Pacific region.

WANG GUANZHONG is a lieutenant general in and deputy chief of staff of China's People's Liberation Army. He is also a member of the Chinese Communist Party's Central Committee.

EXPLORING THE ISSUE

Is China Being the Aggressor in the Issue of the Disputed Island in the Pacific Ocean?

Critical Thinking and Reflection

1. Within the United States and most other countries, land claim disputes would be submitted to and adjudicated by the country's courts. Moreover, even disappointed claimants would have to abide by the courts' decisions. If losing parties did not, the countries' law enforcement authorities would enforce the decisions. For the most part, that is not how things work in international affairs. There are no central enforcement authorities, and what international courts there are have limited jurisdiction. The United States and many other countries, for example, either do not recognize the jurisdiction of the UN's International Court of Justice (ICJ) or reserve the right to ignore its rulings. Would the world work better if all countries granted authority to the ICJ and agreed to abide by its decisions, whether favorable or not? Would you support that position for your own country?

2. China probably has the military power to control most of the areas it claims, except those islands in dispute with Japan. Beyond that the United States is the only country capable of countering Chinese forces in the disputed regions. Would you support U.S. naval action to defend the disputed areas? What U.S. interests are at stake?

3. Would it be wise for the United States to try to restrain China's power and/or build alliances, weapons systems, and other measures to counterbalance it by supplying weapons and support to other counties in the region and by encouraging Japan to significantly rearm?

Is There Common Ground?

There is common ground, but it will be difficult to find. One of many difficulties is that the disputes are rooted in part in national pride. This is especially true for China. Like Russia and other great powers that have fallen on hard times in the past, recovery brings with it an urge to reclaim the status of global or regional power. Part of that is the frequent assertion of a sphere of influence over nearby areas. This is akin to the Monroe Doctrine, which since 1823 has asserted "as a principle . . . the rights and interests of the United States" in the rest of the Western Hemisphere. Persuading China to be an "equal" in Asia contradicts the urge of great powers to claim their sphere of influence. A second problem is that some of the disputed areas have or are thought to have important resources. Oil is most notable, but natural gas, various minerals, and fish stocks are also important. Third, as noted in the critical thinking section, there is no widely accepted international dispute resolution mechanism, such as a court system, operating in the world today, nor is there an effective enforcement capability or agency. In the end, some accommodation may be reached because the countervailing military forces and the economic, human, and other costs of war deter China and the other claimants from using force. The United States and China, for instance, have huge sums invested in one another through ownership of each other's companies, real estate, stocks, government and private bods, and other investment vehicles. China's cumulative investment in the United States was $1.7 trillion in 2013, and Americans have hundreds of billions of dollars invested in China. Adding to those sums, China–U.S. annual trade in goods and services in 2014 was well over a half trillion dollars. Warfare between the two economic behemoths would be a financial disaster for both no matter who prevailed.

But international relations are not driven fully by logic, and a "roll of the iron dice" could occur. That is a phrase used by German Chancellor Theobald von Bethmann-Hollweg in 1914 to describe the gamble Germany was taking in 1914 during a crisis in Europe. It was a poor bet, and the chancellor crapped out. World War I ensued; 16.5 million soldiers and civilians died; Germany was defeated.

Create Central

www.mhhe.com/createcentral

Additional Resources

Shambaugh, David (2014). *China Goes Global: The Partial Power* (Oxford University Press)

One of the several capable introductory studies that would be helpful.

Scobell, Andrew, and Nathan, Andrew J. (October 2012). "China's Overstretched Military," *The Washington Quarterly* (35/4: 135–148)

A less worrisome view of China's military than given by Wortzel below.

Wortzel, Larry M. 2013). *The Dragon Extends Its Reach: Chinese Military Power Goes Global* (Potomac Books)

The title explains Wortzel's point of view.

Zhang, Biwu (2012). *Chinese Perceptions of the U.S.: An Exploration of China's Foreign Policy Motivations* (Lexington Books)

A recent study of China's views of the United States.

Internet References . . .

China's Maritime Disputes (2013)

A good overview of the maritime regions in dispute including some video material is available of the website of the Council on Foreign Relations.

www.cfr.org/asia-and-pacific/chinas-maritime-disputes/p31345#!/.

"Q&A: South China Sea Dispute"

An informative background article from May 2014 by the BBC.

www.bbc.com/news/world-asia-pacific-13748349 (May 8, 2014)

Shangri-La Dialogue

For more on the conference series, go to the website of the International Institute for Strategic Studies. Particularly note the keynote address given by Prime Minister Abe of Japan and how it contrasts with the address of Secretary of Defense Hagel. Both take China to task, but Abe does so elliptically, while Hagel is blunt.

www.iiss.org/en/events/shangri-s-la-s-dialogue

Unit 2

UNIT

Middle East Issues

*O*f all the world's regions, the Middle East is currently the world's most troubled. Some of the problems date back to biblical times. The Bible/Torah opens with the Jewish people suffering under slavery in Egypt. They escape under the leadership of Moses, but their search for a homeland brings them into conflict with other groups in and around what is now Israel. Defeat by the Romans and other factors drive most Jews out of the area, but then beginning in the early 20th century many Jews from Europe and elsewhere move to the traditional Jewish homeland and Israel is reborn as a country after World War II. Once again, however, that process occurs amid violence with others in the region, particularly the Palestinians residing in the area claimed by the Jewish Israelis. By this point and since the late 6th century, most Palestinian and other neighbors are Muslims. The mixture of nationality and religion between the two groups makes tensions even sharper.

Tensions in the region are not just inter-religious. They are also intra-religious. Muslims are split into two major groups, the Sunnis and the Shi'ites, and there are many subgroups in each. There is also a split in the region between religious traditionalist/fundamentalists and more secular Muslims. Regional rivalries are also the result of basic nationalism, with borders, resources, and other factors leading to strife. All of this would be troubling enough, but the region's vast oil wealth makes its control and/or defense a major concern for the rest of the energy-hungry world. Thus instability in the area often invites foreign intervention.

Selected, Edited, and with Issue Framing Material by:
John T. Rourke, *University of Connecticut, Storrs*

ISSUE

Is Chaos in the Middle East Largely the Fault of U.S. Policy?

YES: Elliott Abrams, from "The Man Who Broke the Middle East," *The Politico* (2014)

NO: Jeffrey Goldberg, from "No, President Obama Did Not Break the Middle East," *The Atlantic* (2014)

Learning Outcomes

After reading this issue, you will be able to:

- Identify all the major points of conflict in the Middle East.
- Give an overview of the various doctrinal divisions among and within the nationalities, religions, and cultures of the people living in the Middle East.
- Have an informed opinion about impact U.S. policy has had on the Middle East during the Obama administration.

ISSUE SUMMARY

YES: Elliott Abrams, a senior fellow for Middle Eastern studies at the Council on Foreign Relations and a former top foreign policy official under President Ronald Reagan and President George W. Bush, tells readers that President Barack Obama's policy throughout the Middle East has been a story of failure and danger. He argues that the Middle East that Obama inherited in 2009 was largely at peace and that Obama's policies have actively or passively occasioned the regions to become tumultuous and dangerous.

NO: Jeffrey Goldberg, a national correspondent for *The Atlantic* specializing in the Middle East, writes that Elliott Abrams errs both by overstating the degree to which the Middle East was stable at the beginning of the Obama presidency and by laying too much blame on Obama's policies for the current chaotic state of the region.

Perhaps the most challenging part of creating this edition of *Taking Sides* is the task here of writing an introduction to the ensuing debate over Middle East policy that gives a sufficient background to the issue while containing that introduction to two pages or so.

We can start with a few basics. Like most regions, the boundaries of the Middle East are somewhat divergently defined, but most agree that the region certainly includes all the countries from Iraq in the east to Tunisia and perhaps Algeria in the west. Except for Israel and Lebanon, a large majority of the people of these countries are Arabs, speak Arabic, and are Muslims. It is important to note,

though, that most Muslims are not Arabs. Indeed, there are more Muslims in Indonesia, the largest predominantly Muslim country, than there are in the entire Middle East.

Muslims are adherents of Islam. They worship a single deity, Allah, and revere Muhammad (c. 570–632) as a prophet who received Allah's teachings in a vision and had them transcribed into the Quran (Koran). But just as "Christian" is, "Muslim" is an overarching term. Islam includes different groups (faiths, denominations, sects). The most significant division is between the majority Sunnis and the Shiite (Shia) minority. The issues between the two sects involve doctrinal matters beyond our inquiry. What is important here is that the sometimes quiescent

Sunni–Shiite rivalry was reignited in 1979 when the Ayatollah Khomeini led fundamentalist Shiites to power in Iran. One result was a long war with Iraq (1980–1988). One cause was Khomeini's determination to overthrow the Sunni-dominated regime of Saddam Hussein. Beyond this division, there are many others in Islam. For example, the Alawites are a Shia sect that are a minority in mostly Sunni Syria but dominate its government because President Bashar al-Assad is Alawite.

Islam, again like Christianity, is also divided between traditionalists (fundamentalists) and secularists. Fundamentalists do not believe in the separation of mosque and state. Instead they believe that Muslims should be governed by religious leaders under Islamic religious law (the sharia) interpreted by religious courts. This concept is why Iran's grand Ayatollah is more important politically than its president. Traditionalists also are very conservative socially. They want to preserve their cultural traditions, such as having women cover their faces in public. Some fundamentalists also want to create a Caliphate, a unified political-religious community (*ummah*) of Muslims everywhere governed by Islamic law and a Caliph as successor to Muhammad. That is the goal of the ISIS—the Islamic State in (greater) Syria—the fundamentalist Sunni movement fighting in eastern Syria and western Iraq in 2014. Secularists, by comparison, believe that within Islam there can be many Muslim countries and that religious and secular law should be kept separate.

Ethno-nationalism also divides the Middle East. Some of this involves rivalries between countries, including the aforementioned determination of Iraq and Iran to dominate the Persian Gulf region. Other ethno-national divisions are internal. For one, Iraq is divided among Sunnis, Shiites, and Kurds. This last group, which is also found in Turkey, Iran, and elsewhere, is not ethnic Arab, speaks its own language, and practices its own form of Sunni Islam.

The history of the Middle East is also part of the instability in the Middle East and to the animosity some Muslims have to the West, including the United States. Europe's Christian kings waged eight crusades against Muslims beginning in 1195. Christianity's Orthodox emperors of Byzantium and later the Orthodox czars of Russia also clashed with Muslims. Over time, Muslim secular strength declined, and the last vestige of Muslim power was eclipsed when the Ottoman Empire centered in Turkey collapsed after World War I. During the war the British had promised Arabs independence in return for support, which the Arabs provided. But simultaneously, the British and French betrayed the Arabs and agreed to divide the Middle East into virtual colonies. This experience of betrayal and colonial domination add to many Arabs' sense of being beset by the largely Christian-heritage West. During the last half century, direct political domination ended, but there is a strong sense among many Muslims that Western dominance, led by the United States, has continued through alleged neocolonialist practices such as protecting authoritarian pro-Western regimes in Saudi Arabia and Kuwait and using military force to smite Muslim countries that defy the United States and its European allies, and by favoring and protecting Israel.

YES ↵

Elliott Abrams

The Man Who Broke the Middle East

There's always Tunisia. Amid the smoking ruins of the Middle East, there is that one encouraging success story. But unfortunately for the [Barack] Obama narratives, the president had about as much as to do with Tunisia's turn toward democracy as he did with the World Cup rankings. Where administration policy has had an impact, the story is one of failure and danger.

The Middle East that Obama inherited in 2009 was largely at peace, for the surge in Iraq had beaten down the al Qaeda-linked groups. U.S. relations with traditional allies in the Gulf, Jordan, Israel and Egypt were very good. Iran was contained, its Revolutionary Guard forces at home. Today, terrorism has metastasized in Syria and Iraq, Jordan is at risk, the humanitarian toll is staggering, terrorist groups are growing fast and relations with U.S. allies are strained.

How did it happen? Begin with hubris: The new president told the world, in his Cairo speech in June 2009, that he had special expertise in understanding the entire world of Islam—knowledge "rooted in my own experience" because "I have known Islam on three continents before coming to the region where it was first revealed." But President Obama wasn't speaking that day in an imaginary location called "the world of Islam"; he was in Cairo, in the Arab Middle East, in a place where nothing counted more than power. "As a boy," Obama told his listeners, "I spent several years in Indonesia and heard the call of the azaan [Islamic call to worship] at the break of dawn and the fall of dusk." Nice touch, but Arab rulers were more interested in knowing whether as a man he heard the approaching sound of gunfire, saw the growing threat of al Qaeda from the Maghreb [northwest Africa] to the Arabian Peninsula, and understood the ambitions of the ayatollahs as Iran moved closer and closer to a bomb.

Obama began with the view that there was no issue in the Middle East more central than the Israeli-Palestinian conflict. Five years later he has lost the confidence of both Israeli and Palestinian leaders, and watched his second secretary of state squander endless efforts in a doomed quest for a comprehensive peace. Obama embittered relations with America's closest ally in the region and achieved nothing whatsoever in the "peace process." The end result in the summer of 2014 is to see the Palestinian Authority turn to a deal with Hamas for new elections that—if they are held, which admittedly is unlikely—would usher the terrorist group into a power-sharing deal. This is not progress.

The most populous Arab country is Egypt, where Obama stuck too long with [President] Hosni Mubarak as the Arab Spring arrived, and then with the Army, and then the Muslim Brotherhood President Mohammed Morsi, and now is embracing the Army again. Minor failings like the persecution of newspaper editors and leaders of American-backed NGOs [nongovernmental organizations: private organization working across borders], or the jailing of anyone critical of the powers-that-be at a given moment, were glossed over. When the Army removed an elected president [Morsi], that was not really a "coup"—remember? And as the worm turned, we managed to offend every actor on Egypt's political stage, from the military to the Islamists to the secular democratic activists. Who trusts us now on the Egyptian political scene? No one.

But these errors are minor when compared to those in Iraq and Syria. When the peaceful uprising against President Bashar al-Assad [in Syria] was brutally crushed, Obama said Assad must go; when Assad used sarin gas [a nerve gas], Obama said this was intolerable and crossed a red line. But behind these words there was no American power, and speeches are cheap in the Middle East. Despite the urgings of all his top advisers (using the term loosely; he seems to ignore their advice)—[Director Leon] Panetta at CIA and then Defense, [Secretary Hillary] Clinton at State, [former general and Director David] Petraeus at CIA, even [Chairman of the Joint Chiefs of Staff, Martin] Dempsey at the Pentagon—the president refused to give meaningful assistance to the Syrian nationalist rebels. Assistance was announced in June 2013 and then again in June 2014 (in the president's West Point speech) but it is a minimal effort, far too small to match the presence of

Hezbollah and Iranian Quds Force [part of Iran's Revolutionary Guard] fighters in Syria. Arabs see this as a proxy war with Iran, but in the White House the key desire is to put all those nasty Middle Eastern wars behind us. So in the Middle East American power became a mirage, something no one could find—something enemies did not fear and allies could not count on.

The humanitarian result has been tragic: At least 160,000 killed in Syria, perhaps eight million displaced. More than a million Syrian refugees [are now] in Lebanon (a country of four million people, before Obama added those Syrians), [and] about a million and a quarter Syrian refugees [are] in Jordan (population six million before Obama). Poison gas back on the world scene as a tolerated weapon, with Assad using chlorine gas systematically in "barrel bombs" this year and paying no price whatsoever for this and for his repeated attacks on civilian targets. Both of the key officials handling Syria for Obama—State Department special envoy Fred Hof and Ambassador Robert Ford—resigned in disgust when they could no longer defend Obama's hands-off policy. Can [U.S. Ambassador to the United Nations] Samantha Power be far behind, watching the mass killings and seeing her president respond to them with rhetoric?

The result in security terms is even worse: the largest gathering of jihadis we have ever seen, 12,000 now and expanding. They come from all over the world, a jihadi Arab League, a jihadi EU, a jihadi U.N. Two or three thousand are from Europe, and an estimated 70 from the United States. When they go home, some no doubt disillusioned but many committed, experienced and well trained, "home" will be Milwaukee and Manchester and Marseille—and, as we see now on the front pages, to Mosul. When Obama took office there was no such phenomenon; it is his creation, the result of his passivity in Syria while Sunnis were being slaughtered by the Assad regime.

And now they have spread back into Iraq in sufficient numbers to threaten the survival of its government. Obama has reacted, sending 300 advisers, a number that may presage further expansion of American military efforts. Perhaps they will find good targets, and be the basis for American air strikes and additional diplomatic pressure. But we had won this game, at great expense, before Obama walked away. The fiery rage of Iraqi Sunnis at the government in Baghdad had been banked by 2009. American diplomatic efforts, whose power was based in the American military role, disappeared under Obama, who just wanted out. It was his main campaign pledge. So we got out, fully, completely, cleanly—unless you ask about the real world of Iraq instead of the imaginary world of campaign speeches. We could no longer play the role

we had played in greasing relations between Kurds, Shia and Sunnis, and in constraining Prime Minister Nouri al-Maliki's [a Sunni] sectarian excesses. The result was an Iraq spinning downward into the kind of Sunni-Shia confrontation we had paid so dearly to stop in 2007 and 2008, and ISIS—the newest moniker for al Qaeda in Iraq—saw its chance, and took it. [ISIS: Islamic State in Iraq and al-Sham. Al-Sham refers to the Levant, an area encompassing Syria, Lebanon, Israel, and Jordan.]

So now we're back in Iraq—or maybe not. Three hundred isn't a very large number; it is instead reminiscent of the 600 soldiers Obama sent to Central and Eastern Europe after the Russians grabbed Crimea and started a war in Ukraine. Who is reassured by that number, 600, and who is scared by it? Same question for Iraq: Are the Gulf allies reassured by "up to 300" advisers? Is General Qassem Soleimani, the dark mastermind of the Iranian Revolutionary Guards, quaking now?

If there is one achievement of Obama policy in the Middle East (because Tunisia's genuine success isn't America's to claim) it is to advance reconciliation between Israel and the Gulf states. This will not be celebrated by the White House, however, because they are joined mostly in fear and contempt for American policy, but it is an interesting development nonetheless. If there is one thing the Gulf Sunni kingdoms understand, it is power—in this case, the Iranian power they fear (as they once feared Saddam's power, and were saved from it by America). The king of Jordan incautiously spoke several years ago about a "Shia crescent," but even he must have thought it would take far longer to develop. A map that starts with Hezbollah [a Shia group] in Beirut's southern suburbs and traces lines through Syria and Iraq into [Shia-dominated] Iran would now not be just a nightmare vision, but an actual accounting of where Iran's forces and allies and sphere of influence lie. That's what the Saudis, Emiratis, Kuwaitis and others see around them, growing year by year while their former protector dithers. They see one other country that "gets it," sees the dangers the same way, understands Iran's grasp at hegemony just as they do: Israel. Oh to be a fly on the wall at the secret chats among Sunni Gulf security officials and their Israeli counterparts, which must be taking place in London and Zurich and other safe European capitals. In the world they all inhabit the weak disappear, and the strong survive and rule. They are the ultimate realists, and they do not call what they see in Washington "realpolitik."

From World War II, or at least from the day [in 1963 when] the British left Aden [a seaport in Yemen], the United States has been the dominant power in the Middle East. Harry Truman backed the Zionists and Israel came into being; we opposed Suez [the British-French-Israeli seizure of the Suez Canal in 1956] so the British, French and Israelis backed off; we became the key arms supplier for all our friends and kept the Soviets out; we reversed Saddam's grabbing of Kuwait; we drove him from power; we drew a red line against chemical warfare; we said an Iranian bomb was unacceptable.

But that red line then disappeared in a last-minute reversal by the president that to this day is mentioned in every conversation about security in the Middle East, and no Arab or Israeli leader now trusts that the United States will stop the Iranian bomb. After all, we have passively watched al Qaeda become a major force in the heart of the region, and watched Iran creep closer to a nuclear weapon, and watched Iran send expeditionary forces to Syria—unopposed by any serious American pushback. Today no one in the Middle East knows what the rulebook is and whether the Americans will enforce any rules at all. No one can safely tell you what the borders of Iraq or Syria will be a few years hence. No one can tell you whether American power is to be feared, or can safely be derided.

That's the net effect of five and a half years of Obama policy. And, to repeat, it is Obama policy: not the collective wisdom of Kerry and Clinton and Panetta and Petraeus and other "advisers," but the very personal set of decisions by the one true policymaker, the man who came to office thinking he had a special insight into the entire world of Islam. In the Middle East today, the "call of the azaan" is as widely heard as Obama remembered from Indonesia. But when leaders look around they see clever, well-resourced challenges from Shia and Sunni extremists armed to the teeth, with endless ambitions, willing to kill and kill to grasp power—and far more powerful today than the day this president came into office. They do not see an American leader who fully understands those challenges and who realizes that power, not speeches, must be used to defend our friends and allies and interests. So there's one other thing a lot of Israeli and Arab leaders share, as they shake their heads and compare notes in those secret meetings: an urgent wish that Jan. 20, 2017, were a lot closer.

ELLIOTT ABRAMS is senior fellow for Middle Eastern studies at the Council on Foreign Relations in Washington, D.C. He served as a deputy national security adviser to the president (2005–2009) and as assistant secretary of state for inter-American affairs (1985–1989). He has a JD degree from Harvard University.

Jeffrey Goldberg  **NO**

No, President Obama Did Not Break the Middle East

A brief note on a new Elliott Abrams essay in *Politico Magazine* that appears under the eye-catching headline, "The Man Who Broke the Middle East." The man in question is not Sykes, or Picot or [Egyptian President Gamel Abdel] Nasser [1956–1970], or Saddam [Hussein, president of Iraq, 1979–2003], or [Ruhollah Mostafavi Moosavi] Khomeini [Iran's Grand Ayatollah and leader, 1979–1989] or George W. Bush or [Iraq's Prime Minister] Nouri al-Maliki, but Barack Obama. [The Middle East territories of the near defunct Ottoman Empire were divided after its defeat in World War I by an agreement negotiated by British diplomat Sir Mark Sykes and French diplomat François Georges-Picot. The boundaries of the resulting colonies and areas of influence for the two countries closely resemble the current states in the region.] I often agree with Elliott, but I could not let this one go by without a response. Don't worry. This won't take long.

Here is Elliott's thesis:

> The Middle East that Obama inherited in 2009 was largely at peace, for the surge in Iraq had beaten down the al Qaeda-linked groups. U.S. relations with traditional allies in the Gulf, Jordan, Israel and Egypt were very good. Iran was contained, its Revolutionary Guard forces at home. Today, terrorism has metastasized in Syria and Iraq, Jordan is at risk, the humanitarian toll is staggering, terrorist groups are growing fast and relations with U.S. allies are strained.

A few points. The first is to note that the Middle East Obama inherited in early 2009 was literally at war—Israel and the Gaza-based Hamas were going at each other hard until nearly the day of Obama's inauguration. Obama managed to extract himself from that one without breaking the Middle East.

In reference to a "contained" Iran, I would only note that Iran in 2009 was moving steadily toward nuclearization, and nothing that the Bush administration, in which Elliott served, had done seemed to be slowing Iran down. Flash forward to today—the Obama administration (with huge help from Congress) implemented a set of sanctions so punishing that it forced Iran into negotiations. (Obama, it should be said, did a very good job bringing allies on board with this program.) Iran's nuclear program is currently frozen. The Bush administration never managed to freeze Iran's nuclear apparatus in place. I'm not optimistic about the prospects for success in these negotiations (neither is Obama), but the president should get credit for leading a campaign that gave a negotiated solution to the nuclear question a fighting chance.

It's also worth noting that when Obama came to power, he discovered that the Bush administration had done no detailed thinking about ways to confront Iran, either militarily or through negotiations. There was rhetoric, but no actual planning. Obama applied himself to this problem in ways that Bush simply did not.

Elliott writes that, in 2009, U.S. relations with Arab allies were good. But these relations, in many cases, were built on lies and morally dubious accommodations. He states that "the most populous Arab country is Egypt, where Obama stuck too long with [Egyptian President] Hosni Mubarak as the Arab Spring arrived, and then with the Army, and then the Muslim Brotherhood President Mohammed Morsi, and now is embracing the Army again."

Let's break this down for a minute. It was the policy of several administrations to maintain close relations with Egypt's military rulers. It was Bush administration policy to maintain close relations with Mubarak. Perhaps the 2011 uprising in Egypt could have been avoided had the Bush administration, in honoring its "Freedom Doctrine," engineered Mubarak's smooth departure several years before Cairo exploded. Obama inherited a dysfunctional relationship with Egypt from his predecessor. This is not to excuse the administration's faltering and sometimes contradictory approach to the Egypt problem today, but simply to set it in some context.

On the peace process, Elliott writes,

Obama began with the view that there was no issue in the Middle East more central than the Israeli-Palestinian conflict. Five years later he has lost the confidence of both Israeli and Palestinian leaders, and watched his second secretary of state squander endless efforts in a doomed quest for a comprehensive peace. Obama embittered relations with America's closest ally in the region and achieved nothing whatsoever in the "peace process." The end result in the summer of 2014 is to see the Palestinian Authority turn to a deal with Hamas for new elections that—if they are held, which admittedly is unlikely—would usher the terrorist group into a power-sharing deal. This is not progress.

I'm sure Elliott remembers that in 2006, the Bush administration helped bring the terrorist group Hamas to power, by engineering elections that neither the Palestinian Authority nor Israel actually wanted. I'm sure he also remembers that President Bush (along with a series of presidents before him) failed utterly to bring about a peace treaty between Israel and the Palestinians. It seems a bit unfair to single-out Obama for failing at something presidents of both parties, for 40 years, have also failed to accomplish.

On Syria and Iraq, Elliott is on somewhat firmer ground. I've argued that an earlier intervention in Syria, in the form of support for what was then a more-moderate rebel coalition, might have changed the balance of power.

On a deeper level, the idea of blaming any American president for the terrible state of the Middle East seems somewhat dubious. I argue this question with myself and with my friends all the time, because I do recognize that the U.S. has a singular role to play in the world's most volatile and dysfunctional region, and I agree with [scholar] Robert Kagan, who argues that superpowers don't have the luxury of taking vacations from responsibility. But on the other hand, conditions in Iraq, while aggravated by certain Obama policies, cannot be pinned on him alone. For that matter, the man who truly broke Iraq was not George W. Bush, but Saddam Hussein, who through murder, rape, pillage, torture, and genocide destroyed millions of Iraqi lives.

Jeffrey Goldberg is a national correspondent for *The Atlantic* specializing in the Middle East and a recipient of the National Magazine Award for Reporting. He attended the University of Pennsylvania.

EXPLORING THE ISSUE

Is Chaos in the Middle East Largely the Fault of U.S. Policy?

Critical Thinking and Reflection

1. Apart from the vast reserves of oil and natural gas located in the Middle East, what are the U.S. national interests in that region?
2. Which should be more important in determining the U.S. position on the possible transition of any government or type of government of another country: whether the new government will be democratic or whether it will be friendly toward the United States?
3. What, if any circumstances, would cause you to support the intervention of U.S. ground troops in the Middle East?

Is There Common Ground?

The "blame game," which Elliot Abrams certainly engages in, usually leaves little room for common ground. It is, however, important to try to understand the "whys" behind the ongoing, multifaceted trauma in the Middle East. There is an old warning about trying to fix what isn't broken, but a less well-known but equally wise bit of advice is that you cannot fix something unless you know why it is broken. Understanding why can lead you to choose the right policy tools to improve a situation. It can also lead to the conclusion that there may not be anything you can do, no matter how stressful a situation is. Knowing what has caused a problem and what is required to fix it also brings up the question of what price you are willing to pay. Could U.S. forces oust President Assad in Syria, prevent ISIS fighters from holding the ground they have taken, keep Iraq from disintegrating, and keep Libya from imploding? Yes, to a substantial degree, but only so long as Americans remained in place and at financial cost and with American dead and wounded.

There is also the almost always unanswerable question of how intervention will impact the future. If the United States, for instance, had not toppled Saddam Hussein or supported rebels trying to oust President Assad, would ISIS be committing the abominations it is currently inflicting in large parts of Iraq and Syria? Perhaps Star Trek's "prime directive" is right and, as Captain Jean-Luc Picard put it, "no matter how well intentioned that interference may be, the results are invariably disastrous."

Create Central

www.mhhe.com/createcentral

Additional Resources

Fisher, Max. (May 5, 2014). "40 Maps That Explain the Middle East," *Vox* online at www.vox.com/a /maps-explain-the-middle-east

Maps are an often-underappreciated way of understanding history and current events. A fine source using maps is Fisher's contribution to *Vox*. Pause at each map and note how they change to show the progression of events.

Looney, Robert, ed. (2014). *Handbook of US-Middle East Relations* (Routledge)

Migdais, Joel S. (2014). *Shifting Sands: The United States in the Middle East* (Columbia University Press)

Internet References . . .

Best of History—Middle East

www.besthistorysites.net/index.php/modern-history
/middle-east-conflict

Middle East—Al Jazeera

www.aljazeera.com/news/middleeast/

Middle East—BBC.com

www.bbc.com/news/world/middle_east/

White House—Middle East and Africa

http://www.whitehouse.gov/issues/foreign
-policy#section-middle-east-and-north-africa

Unit 3

UNIT

Economic Issues

*I*nternational economic issues have an immediate and personal effect on individuals in ways that few other types of issues do. These issues include such factors as foreign trade, the flow of investment capital across borders, the distribution of flow of oil and other vital resources internationally, and the relative values (exchange rate) of currencies. Because each country's strength, absolutely and relatively, is based in significant part on its economic strength, domestic economic issues and policy are also part of the global equation.

International economics has always been a factor in world affairs, but it has become even more central in recent times. Trade, investment, and monetary exchange have all skyrocketed in the last century. As a result, the intertwining of countries economically has proceeded to the point that it is referred to as globalization. This phenomenon also has cultural, environmental, and other aspects.

Another change is the spread of capitalism, at least restrained capitalism, as the dominant economic model. Communism as a model is near dead, and socialism is teetering. Whether this change is good, and for whom, is more debatable.

Economic cooperation is not always the norm, however. Rivalries for resources and other factors cause conflict. Countries also use their economic resources as instruments of their foreign policy. One recent example is the economic sanctions the United States and, to a lesser degree, Western Europe have levied against Russia for its actions in Ukraine. One reason Western Europe has not responded as strongly as the United States to Russia's actions is that the Europeans depend on natural gas from Russia and are wary of Moscow's threats to turn off that energy supply in retaliation for sanctions.

Selected, Edited, and with Issue Framing Material by:
John T. Rourke, *University of Connecticut, Storrs*

ISSUE

Will the European Union Collapse?

YES: Guy Millière, from "The Coming Collapse of the European Union," *Gatestone Institute* (2013)

NO: José Manuel Barroso, from "The Annual 'State of the [European] Union' Address," *European Parliament, Strasbourg, France* (2013)

Learning Outcomes
After reading this issue, you will be able to:
• Explain briefly what the European Union is and how it has evolved.
• Identify the prominent contemporary issues that lead some to doubt the future of the European Union.
• Have some early sense of whether you think the model of the European Union would be a good one for the integration of the countries, including the United States and Canada, in North America or even the entire Western Hemisphere.

ISSUE SUMMARY

YES: Guy Millière, a senior fellow of the Gatestone Institute, a private research and advocacy organization based in New York City, writes that creating the European Union (EU) was an effort to build a society based on abstract principles—a united and peaceful Europe—without considering historical, social, and economic realities, as if its members were infinitely malleable. Millière argues that such efforts have often led to disaster, and that the European Union is no different.

NO: José Manuel Barroso, president of the European Commission for the European Union, tells members of the European Parliament that the global economic recession that began in 2008 and other problems have severely tested the EU in recent years and continue to do so. But he also contends that the members of the EU are tackling their challenges together and will overcome them, and that the EU will not only survive, but become stronger.

In the aftermath of World War II (1939–1945), the shattered countries of Europe began the process of diminishing the importance of national borders and identities and building common institutions and identities as a way to ensure future peace and prosperity. The first step came in 1952 when Belgium, France, (West) Germany, Italy, Luxembourg, and the Netherlands created the European Coal and Steel Community, a common market for coal, iron, and steel products. From this genesis, the European community has increased in membership, has expanded its functions, and has undergone a series of name changes. The current name, European Union (EU), was adopted in

1993 by the then 12 members of the organization to symbolize the EU's goal of becoming a single economic entity. The EU now includes 28 countries with a combined population of over 506 million, and a gross domestic product (2013, GDP) of $17.4 trillion, somewhat larger that year than that of the United States ($16.8 trillion). In addition to very advanced economic integration, the EU has also achieved significant political integration. It has its own executive branch, parliament, and courts, and its annual budget in 2014 was approximately $187 billion.

Another key step toward economic integration of the EU was taken in 1993 when the EU countries agreed to eventually give up their national currencies and adopt a single common currency, the euro. Not all countries had to adopt the euro, and some such as Great Britain did not do that. Those countries that chose to adopt the euro had to commit to meeting certain standards for prudent government finance such as keeping their budget deficits to a low level. The countries that use the euro are referred to as the eurozone. As of 2014, there were 18 countries in the eurozone. They included Germany, France, and all the other major economies in the EU other than Great Britain. The euro began to be used for transactions between eurozone countries in 1999 and went into general circulation for all transactions within and between eurozone countries in 2002.

For a number of years, the euro worked well. A single currency made economic transactions between the eurozone countries much easier, and the currency maintained its strength on the international monetary exchange markets. But there were problems. A key one was that the some eurozone countries did not abide by criteria for eurozone membership. Just as was occurring in the United States, budget deficits soared. The eurozone's problems were made markedly worse, such as the U.S. economy was, by the global recession that took hold in late 2008 and that continues to plague government finance around the world.

The crisis created something of a north–south split in the EU. Most of the countries in economic trouble are in the south, with Greece, Spain, Italy, and Portugal the worst off. In the north are Germany and most of the other eurozone countries with the financial resources to assist the trouble countries in avoiding defaulting on their debt and making Europe's shaky economic situation worse. The northern countries have insisted that bail-out money to Greece and the other countries in trouble be contingent on those countries agreeing to changes designed to reduce their budget deficits. The troubled countries have resisted these requirements because they included reduced services, higher taxes, and other politically unpopular policies. Greece's government fell in early 2012 over public anger at its attempt to introduce austerity measures. Since then, Greece has undertaken some reforms and its finances are in somewhat better shape, but the north–south divide remains. Unemployment rates in 2014, for example, were 27 percent in Greece and 26 percent in

Spain compared to 5 percent in Germany and 11 percent for the EU overall.

Beyond this specific crisis, support for the EU has declined somewhat among people in its member countries. In 1990, the Eurobaromoter's polls found 58 percent saying that their country had benefited from EU membership. By 2012, that sense of benefit had declined to 52 percent, while the not benefited response had risen from 23 to 37 percent. One issue is what many Europeans consider to be the overextension of the EU's bureaucracy, the European Commission, which the author of the NO selection, José Manuel Barroso, heads. In 2000, 45 percent of survey respondents said they tended to trust the European Commission; that trust declined to 32 percent in 2014. It should be noted, though, there are also positive signs about the EU in public opinion. In the aftermath of the economic downturn in 2008, those who were hopeful and discouraged about the EU's future were closely balanced, with 49 percent optimistic and 46 percent pessimistic. By mid-2014, 56 percent were optimistic, and the share of pessimists had declined to 38 percent. There is also positive news for the EU in terms of its citizens' desires for greater integration. On a one (slowest) to seven (fastest) scale), people have consistently favored speeding up the pace of integration over what it is. In 1995, the average respondent estimated the then current speed at 3.5 and wished it was 4.8. The year 2013 saw the perceived speed at 3.2 and the preferred speed at 5.0.

There is also a range of political challenges facing the European Union. If the Conservative Party and Prime Minister David Cameron remain in power in Great Britain, he has pledged the government will hold a referendum in which voters can decide whether their country will remain part of the EU or withdraw from it. The 2014 crisis over Russia and Ukraine is also testing the EU. It is unclear whether Europe can act with enough unity and resolve to impose the kinds of economic and diplomatic sanctions that have any chance of deterring Russia from adding more Ukrainian territory than the Crimea region that has already been occupied. In the first of following two selections, Guy Millière predicts that a decade from now Europe will be poorer, more divided, and more confrontational, more violent, and that the European Union will look like a shattered dream.

YES

Guy Millière

The Coming Collapse of the European Union

How could so many clever people get it so wrong? The question was recently asked by the British politician Daniel Hannan in an article on the collapse of the euro; in the coming months, the same question will be asked more and more often about Europe itself.

Europe as it has been built may appear at best a huge error, and at worst a crime against the spirit of liberty that was supposed to be the initial source of inspiration for the whole edifice.

The idea that it is possible to build a society based on abstract principles—without considering historical, social and economic realities, as if its members were infinitely malleable—has often led to disaster; this time is no different.

The formation of this error began in the aftermath of World War II. Looking at the ruins left by Nazism and Fascism, politicians from various European countries fabricated a project meant to erase all past mistakes committed on the continent. They only repeated the mistakes.

Believing that nations, identities, and differences were the essential source of troubles, they decided to make a clean sweep and build an entirely new society, inhabited by new human beings modeled from the top down: Europeans.

Realizing that democracy can be dangerous—it had brought to power the likes of Adolph Hitler—they developed institutions to place it under close supervision. Deducing from their readings and their worldview that a strong economy should be a command economy, they introduced instruments of planning in the production of coal and steel, then in agriculture.

This resulted in the creation of supranational structures that were added to each other over time, in a form of cumulative stacking, until, in 1957, a committee [the European Commission] made up of unelected people was established in Brussels and charged with producing regulations for businesses and aspects of everyday life:

the "efficient use" of water or the "appropriate shape" of waste containers, the "normal curvature" of bananas and cucumbers or the "harmonized definition of yoghurt." Meetings of Heads of States and Governments—a group that later became the European Council—were held out of sight to make key decisions , such as subsidies to industries, price-support mechanisms, decisions about enlargement, the reform of institutions, and the extension of powers. A European Parliament was created [1979], but was given no power.

While the structures piled up, the edifice expanded: after the collapse of the Soviet empire, a group that had started with six countries became 27 member countries. Its name changed: formerly known as European Economic Community, it became European Union [EU, in 1993]. While at first it seemed to be an economic edifice, it has proven to be a political edifice.

It might look like a state [a sovereign/independent country], but it is not a state. It is supposed to be composed of independent countries, but countries that are part of it have a sovereignty [independence] that is more and more limited. It now has two presidents, one is the head of the Commission, the other is the head of the Union, but neither is the holder of any executive office. As the parliament essentially has an "advisory" role, the real power is exercised by the members of the Commission, a bureaucracy of technocrats, and by the European Council, a small political elite. [The European Council is composed of the leaders of the EU's member countries and meets at least four times a year].

People still vote in each country [for their national governments and for their members of the European Parliament], preserving the illusion that national governments have a voice, but 70% of the laws and guidelines in today's Europe are issued at levels that are unattainable for national governments. . . .

The introduction of the single currency was designed to move towards a fuller political, fiscal and social unity, and a more complete abolition of the last vestiges of democracy. The unity in question has not emerged. The euro was supposed to be the currency of all Europe; it is the currency of only 17 countries [those in the "eurozone," of the 28 member countries as of August 2014].

Rather than becoming closer, these countries have gradually moved away from each other, and divisions and disparities have intensified, leading to increasing dysfunctions. Neither Greece, Italy, Spain, Portugal, Ireland, nor even France has seen their economies and societies come to resemble German or Scandinavian economies and societies. Differences in productivity have only widened. Interest rates appropriate for some economies have proven inadequate for other economies. The exchange rate of the euro became too high for some countries, and if devaluation was impossible, the only remaining variables able to be adjusted were unemployment, budget deficits, and debts. On behalf of the construction of the new society, massive financial transfers were conducted from richer countries to poorer countries, while people living in the poorer countries enjoyed for a time the illusion of becoming wealthier without having either to work harder or become more efficient and disciplined.

Other factors played their part: The countries of Europe created welfare states that claimed to support people from the cradle to the grave; they developed economic systems of redistribution that have created in people an appetite for passivity and dependence, leading to a growing number of people who think the government owes them a means of subsistence. As welfare rolls and subsidies have increased, the number of those who contribute to them financially has steadily declined. European nations' borrowing power has gradually diminished, and the redistribution systems have been left unfunded. As the birthrate of native Europeans has fallen sharply, its aging population increases even further the financial burden on those who pay out to it. Immigrants from poor countries have flowed in, most of whom are Muslims who have not integrated, and have ended up living in separate enclaves of lawlessness and resentment. These conditions, coupled with economic decline, have created ghettos ever ready to riot and explode.

A collapse is near. Some EU countries are already in a state of bankruptcy, even if it is kept under wraps. Doubts hang over the medium-term strength of the nations who seem in better shape: the interdependencies that have developed appear difficult to break without causing even greater financial problems, and the health of healthier countries is merely relative. Germany's debt is 83% of its gross domestic product, and it has the lowest birthrate in Europe, 1.36 children per woman. Italy, by comparison,

has a debt of 118% and a birthrate of 1.40. [A fertility rate of about 2.1 keeps a population stable; below that rate the population declines.]

The agreement of December 9 [2011], described as the "Last Chance Agreement," will hold for a few months, but probably not more. [The European Fiscal Compact requires EU members to exercise greater budgetary restraints designed to reduce deficits.] Its goal is to give the high European authorities the power to ratify all countries' budgets in the eurozone, an approach that means guardianship by the unelected bureaucracy of technocrats and the political elite. The same causes will produce the same effects: the countries already in bankruptcy will continue to sink because their economies are not viable with, and often without, the current exchange-rate of the euro. The unelected bureaucracy of technocrats and the political elite will ask for sacrifices. The standard of living will continue to fall throughout the entire eurozone, and deficits will continue to widen. The recession in the countries already in bankruptcy will become permanent. As Germany is unable to bail out these countries indefinitely, they will be doomed sooner or later actually to file for bankruptcy. Germany is no longer finding buyers for its debt. Who will purchase Greek or Italian debt? The Financial Stability Mechanism will be fed for some time by the IMF [International Monetary Fund] and its main contributors. What country could be expected to continue endless assistance, given that difficulties appear more and more serious and will not stop?

The stated aim of the agreement is political, fiscal and social unity. The real aim is to give the unelected bureaucracy and the political elite the authority to implement this unity, at all costs. The argument used is always the same: the union of Europe is the only guarantee of peace and prosperity; without it there will be chaos.

The reality may be very different. There could be chaos precisely because of the European union as it was implemented.

In a book [*EUSSR: The Soviet Roots of European Integration*] published a few years ago [in 2004], the former [Soviet] dissident Vladimir Bukovsky compared the former Soviet Union, or USSR, to the European Union, which he called the EUSSR. Immediately prior to the Soviet Union's falling apart, its leaders were trying to clog the holes and persuade themselves that the Union would last. Although the European leaders are also now clogging the holes and trying to persuade themselves that the European Union will last, these measures will not prevent it from falling apart.

No one can know, of course, what Europe will look like in ten years. The only prediction that seems assured is that it will be poorer, more divided, more confrontational, more violent, and that the European Union will look like a shattered dream. The "European dream" as [economist] Jeremy Rifkin called it [in a 2004 book of that name] seems to have crumbled.

GUY MILLIÈRE is a senior fellow of the Gatestone Institute and a professor at the University of Paris. He has published numerous books and articles on politics. He received a doctoral degree from the University of Paris.

José Manuel Barroso  **NO**

Annual State of the [European] Union

In 8 months' time, voters across Europe will judge [during elections for the European Parliament] what we have achieved together in the last 5 years. In these 5 years, Europe has been more present in the lives of citizens than ever before. Europe has been discussed in the coffee houses and popular talk shows all over our continent.

Today, I want to look at what we have done together. At what we have yet to do. And I want to present what I believe are the main ideas for a truly European political debate ahead of next year's elections.

As we speak, exactly 5 years ago [in 2008], the United States government took over Fannie Mae [Federal National Mortgage Association] and Freddie Mac [Federal Home Loan Mortgage Corporation [two government-sponsored enterprises to guarantee and promote home mortgages], bailed out [financial services giant] AIG [American International Group], and Lehman Brothers [another financial services giant] filed for bankruptcy protection.

These events triggered the global financial crisis. It evolved into an unprecedented economic crisis. And it became a social crisis with dramatic consequences for many of our citizens. These events have aggravated the debt problem that still distresses our governments. They have led to an alarming increase in unemployment, especially amongst young people. And they are still holding back our households and our companies.

But Europe has fought back. In those 5 years, we have given a determined response. We suffered the crisis together. We realized we had to fight it together. And we did, and we are doing it.

If we look back and think about what we have done together to unite Europe throughout the crisis, I think it is fair to say that we would never have thought all of this possible 5 years ago.

We are fundamentally reforming the financial sector so that people's savings are safe. We have improved the way governments work together, how they return to sound public finances and modernize their economies. We have mobilized over 700 billion euro to pull crisis-struck countries back from the brink, the biggest effort ever in stabilization between countries.

I still vividly remember my meeting last year with chief economists of many of our leading banks. Most of them were expecting Greece to leave the euro [no longer use euros as its currency]. All of them feared the disintegration of the euro area. Now, we can give a clear reply to those fears: no one has left or has been forced to leave the euro. This year, the European Union enlarged from 27 to 28 member states. Next year the euro area will grow from 17 to 18.

What matters now is what we make of this progress. *Do we talk it up, or talk it down? Do we draw confidence from it to pursue what we have started, or do we belittle the results of our efforts?* [italics in the original]

I just came back from the G20 [a group of 20 countries including the United States with large economies] in Saint Petersburg [Russia]. I can tell you: this year, contrary to recent years, we Europeans did not receive any lessons from other parts of the world on how to address the crisis. We received appreciation and encouragement.

Not because the crisis is over, because it is not over. The resilience of our Union will continue to be tested. But what we are doing creates the confidence that we are overcoming the crisis—provided we are not complacent.

We are tackling our challenges together. We have to tackle them together.

In our world of geo-economic and geopolitical tectonic changes, I believe that only together, as the European Union, we can give our citizens what they aspire: that our values, our interests, our prosperity are protected and promoted in the age of globalization.

So now is the time to rise above purely national issues and parochial interests and to have real progress for Europe: To bring a truly European perspective to the debate with national constituencies.

Now is the time for all those who care about Europe, whatever their political or ideological position, wherever they come from, to speak up for Europe. If we ourselves don't do it, we cannot expect others to do it either.

José Manuel Barroso, State of the [European] Union Address to the European Parliament, January 3, 2012.

We have come a long way since the start of the crisis. In last year's State of the [European] Union speech, I stated that "despite all [our] efforts, our responses have not yet convinced citizens, markets or our international partners."

One year on, the facts tell us that our efforts have started to convince. Overall spreads are coming down. The most vulnerable countries are paying less to borrow. Industrial output is increasing. Market trust is returning. Stock markets are performing well. The business outlook is steadily improving. Consumer confidence is sharply rising.

We see that the countries which are most vulnerable to the crisis and are now doing most to reform their economies, are starting to note positive results. In Spain, as a signal of the very important reforms and increased competitiveness, exports of goods and services now make up 33% of GDP [gross domestic product], more than ever since the introduction of the euro. Ireland has been able to draw money from capital markets since the summer of 2012, the economy is expected to grow for a third consecutive year in 2013 and Irish manufacturing companies are re-hiring staff. In Portugal, the external current account [a measure of money leaving and coming into a country], which was structurally negative [a deficit flow], is now expected to be broadly balanced, and growth is picking up after many quarters in the red. Greece has completed, just in 3 years, a truly remarkable fiscal adjustment, is regaining competitiveness and is nearing for the first time in decades a primary surplus. And Cyprus, that has started the program later, is also implementing it as scheduled, which is the pre-condition for a return to growth.

For Europe, recovery is within sight.

Of course, we need to be vigilant. As Aristotle [384 BC–322 BC] once wrote, "One swallow does not make a summer, nor one fine day." Let us be realistic in the analysis. Let us not overestimate, but let's also not underestimate what has been done. Even one fine quarter doesn't mean we are out of the economic heavy weather. But it does prove we are on the right track. On the basis of the figures and evolutions as we now see them, we have good reason to be confident.

This should push us to keep up our efforts. We owe it to those for whom the recovery is not yet within reach, to those who do not yet profit from positive developments. We owe it to our 26 million unemployed. Especially to the young people who are looking to us to give them hope. Hope and confidence are also part of the economic equation.

If we are where we are today, it is because we have shown the resolve to adapt both our politics and our policies to the lessons drawn from the crisis. And when I say "we," I really mean: "we": it has really been a joint effort.

So let us continue to work together to reform our economies, for growth and jobs, and to adapt our institutional architecture. Only if we do so, we will leave this phase of the crisis behind us as well. There is a lot we can still deliver together, in this Parliament's and this [European] Commission's mandate.

What we can and must do, first and foremost, let's be concrete is delivering the banking union. It is the first and most urgent phase on the way to deepen our economic and monetary union, as mapped out in the Commission's Blueprint presented last autumn. The legislative process on the Single Supervisory Mechanism [to allow the European Central Bank (ECB) to supervise banks in all countries using the euro] is almost completed. The next step is the ECB's independent valuation of banks assets, before it takes up its supervisory role.

Our attention now must urgently turn to the Single Resolution Mechanism. The Commission's proposal is on the table since July and, together, we must do the necessary to have it adopted still during this term. It is the way to ensure that taxpayers are no longer the ones in the front line for paying the price of bank failure. It is the way to make progress in decoupling bank from sovereign risk.

It is the way to remedy one of the most alarming and unacceptable results of the crisis: increased fragmentation of Europe's financial sector and credit markets—even an implicit re-nationalization. And it is also the way to help restoring normal lending to the economy. Because in spite of the accommodating monetary policy, credit is not yet sufficiently flowing to the economy across the euro area. This needs to be addressed resolutely.

Ultimately, this is about one thing: growth, which is necessary to remedy today's most pressing problem: unemployment. The current level of unemployment is economically unsustainable, politically untenable, socially unacceptable. So all of us here in the Commission—and I'm happy to have all my Commissioners today here with me—all of us want to work intensively with you, and with the member states, to deliver as much of our growth agenda as we possibly can, we are mobilizing all instruments, but of course we have to be honest, not all are at European level, some are at national level. I want to focus on implementation of the decisions on youth employment and financing of the real economy. We need to avoid a jobless recovery.

Europe therefore must speed up the pace of structural reforms. At EU level—because there is what can be done at national level and what can be done at European level—the focus should be on what matters most for the real economy: exploiting the full potential of the single market comes first.

We have a well-functioning single market for goods, and we see the economic benefits of that. We need to extend the same formula to other areas: mobility, communications, energy, finance and e-commerce, to name but a few. We have to remove the obstacles that hold back dynamic companies and people. We have to complete connecting Europe.

I'd like to announce that, today, we will formally adopt a proposal that gives a push towards a single market for telecoms. Citizens know that Europe has dramatically brought down their costs for roaming. Our proposal will strengthen guarantees and lower prices for consumers, and present new opportunities for companies. We know that in the future, trade will be more and more digital. Isn't it a paradox that we have an internal market for goods but when it comes to digital market we have 28 national markets? How can we grab all the opportunities of the future that are opened by the digital economy if we don't conclude this internal market?

The same logic applies to the broader digital agenda: it solves real problems and improves daily life for citizens. The strength of Europe's future industrial base depends on how well people and businesses are interconnected. And by properly combining the digital agenda with data protection and the defense of privacy, our European model strengthens the trust of the citizens. Both with respect to internal and external developments, adopting the proposed legislation on data protection is of utmost importance to the European Commission.

The single market is a key lever for competitiveness and employment. Adopting all remaining proposals under the Single Market Act I and II, and implementing the Connecting Europe Facility in the next few months, we lay the foundations for prosperity in the years to come.

We are also adapting to a dynamic transformation on a global scale, so we must encourage this innovative dynamism at a European scale. That is why we must also invest more in innovation, in technology and the role of science. I have great faith in science, in the capacity of the human mind and a creative society to solve its problems. I would like Europe to be leading that effort globally. This is why we—Parliament and Commission—have made such a priority of Horizon 2020 [a research and technological development program] in the discussions on the EU budget.

And whilst fighting climate change, our 20-20-20 goals have set our economy on the path to green growth and resource efficiency, reducing costs and creating jobs. By the end of this year, we will come out with concrete proposals for our energy and climate framework up to 2030. And we will continue to shape the international agenda by fleshing out a comprehensive, legally binding global climate agreement by 2015, with our partners. Europe alone cannot do all the fight for climate change. Frankly, we need the others also on board. At the same time, we will pursue our work on the impact of energy prices on competitiveness and on social cohesion.

We must also pursue our active and assertive trade agenda. It is about linking us closer to growing third markets and guaranteeing our place in the global supply chain. Contrary to perception, where most of our citizens think we are losing in global trade, we have a significant and increasing trade surplus of more than 300 billion euro a year, goods, services, and agriculture. We need to build on that. This too will demand our full attention in the months to come.

And last but not least, we need to step up our game in implementing the European budget. The EU budget is the most concrete lever we have at hand to boost investments. In some of our regions, the European Union budget is the only way to get public investment because they don't have the sources at national level.

Both the European Parliament and the Commission wanted more resources. We have been in that fight together. But even so, one single year's EU budget represents more money—in today's prices—than the whole Marshall Plan [the U.S. aid program to Europe after World War II] in its time!

My point today is clear: together, there is a lot still to achieve before the elections. It is not the time to throw in the towel, it is time to roll up our sleeves.

None of this is easy. These are challenging times, a real stress test for the EU. The path of permanent and profound reform is as demanding as it is unavoidable. Let's make no mistake: there is no way back to business as usual. Some people believe that after this everything will come back as it was before. They are wrong, this crisis is different. This is not a cyclical crisis, but a structural one. We will not come back to the old normal. We have to shape a new normal. We are in a transformative period of history. We have to understand that, and not just say it. But we have to draw all the consequences from that, including in our state of mind, and how we react to the problems.

We see from the first results that it is possible.

And we all know from experience that it is necessary.

It is only natural that, over the last few years, our efforts to overcome the economic crisis have overshadowed everything else.

But our idea of Europe needs to go far beyond the economy. We are much more than a market. The European ideal touches the very foundations of European society.

It is about values, and I underline this word: values. It is based on a firm belief in political, social and economic standards, grounded in our social market economy.

In today's world, the EU level is indispensable to protect these values and standards and promote citizens' rights: from consumer protection to labor rights, from women's rights to respect for minorities, from environmental standards to data protection and privacy.

Whether defending our interests in international trade, securing our energy provision, or restoring people's sense of fairness by fighting tax fraud and tax evasion: only by acting as a Union do we pull our weight at the world stage.

Whether seeking impact for the development and humanitarian aid we give to developing countries, managing our common external borders or seeking to develop in Europe a strong security and defense policy: only by integrating more can we really reach our objectives.

There is no doubt about it. Our internal coherence and international relevance are inextricably linked. Our economic attraction and political traction are fundamentally entwined.

Does anyone seriously believe that, if the euro had collapsed, we or our Member States would still have any credibility left internationally?

Does everyone still realize how enlargement has been a success in terms of healing history's deep scars, establishing democracies where no one had thought it possible? How neighborhood policy was and still is the best way to provide security and prosperity in regions of vital importance for Europe? Where would we be without all of this?

And does everyone still remember just how much Europe has suffered from its wars during the last century, and how European integration was the valid answer?

Next year, it will be one century after the start of the First World War. We must never take peace for granted. We need to recall that it is because of Europe that former enemies now sit around the same table and work together. It is only because they were offered a European perspective that now even Serbia and Kosovo come to an agreement, under mediation of the EU.

Last year's Nobel Peace Prize [which was awarded to the EU] reminded us of that historic achievement: that Europe is a project of peace.

Let me say this to all those who rejoice in Europe's difficulties and who want to roll back our integration and go back to isolation: the pre-integrated Europe of the divisions, the war, the trenches, is not what people desire and deserve. The European continent has never in its history known such a long period of peace as since the creation of the European Community. It is our duty to preserve it and deepen it.

There are those who claim that a weaker Europe would make their country stronger, that Europe is a burden; that they would be better off without it. My reply is clear: we all need a Europe that is united, strong and open.

If you don't like Europe as it is: improve it! Find ways to make it stronger, internally and internationally, and you will have in me the firmest of supporters. Find ways that allow for diversity without creating discriminations, and I will be with you all the way. But don't turn away from it.

There are areas of major importance where Europe must have more integration, more unity. Where only a strong Europe can deliver results. I believe a political union needs to be our political horizon. This is not just the demand of a passionate European. This is the indispensable way forward to consolidate our progress and ensure the future.

So let us work together—for Europe. Let us not forget: one hundred years ago—Europe was sleepwalking into the catastrophe of the war of 1914. Next year, in 2014, I hope Europe will be walking out of the crisis towards a Europe that is more united, stronger and open.

José Manuel Barroso is the president of the European Commission and a former prime minister of Portugal (2002–2004). He holds a law degree from the University of Lisbon.

EXPLORING THE ISSUE

Will the European Union Collapse?

Critical Thinking and Reflection

1. What difference does it make to the United Sates if the European Union were to collapse or, alternately, go on to become a fully integrated "United States of Europe," with most of its citizens feeling primary political loyalty to the EU rather than their individual countries and those countries having authority that is more like the states in the United States rather than like the largely independent countries the EU member countries are now?

2. What do you think of the United States and the rest of the countries of North and South America and the Caribbean joining together to begin to change the current Organization of American States into a future United States of the Americas; adopting binding common policies through such institutions as the Congress of the Americas; and dropping dollars, pesos, escudos, and other national currencies and in their place adopting a common currency, the "continental"?

3. How about a United States of the World?

Is There Common Ground?

Not only is there common ground, it is probable that the future of the EU lies somewhere between Guy Millière's dire portrayal of the future and José Manuel Barroso's rosy prediction.

It seemed to be wishful thinking to the extreme of delusional when in 1943, with World War II raging, French political and economic adviser Jean Monnet first advocated creating a united Europe. As he put it to the French government in exile, "There will be no peace in Europe, if the states are reconstituted [after the war] on the basis of national sovereignty. . . . The countries of Europe are too small to guarantee their peoples the necessary prosperity and social development. The European states must constitute themselves into a federation." Yet eight years later, Belgium, France, Germany, Italy, Luxembourg, and the Netherlands had created the European Coal and Steel Community, and six years after that those six countries signed the Treaty of Rome establishing the European Economic Community (EEC), more commonly known as the "Common Market."

Since then, the number of countries in what has become the EU has more than quadrupled, and its share of the world GDP has expanded by 53 percent from 15 to 23 percent. Perhaps even more remarkably, a substantial degree of political integration now exists. There are myriad EU regulations that bind all members; voters directly elect members of the European Parliament, the European Court of Justice makes decisions that apply to all countries, and with few exceptions citizens of any EU country can freely migrate to and work in any other EU country. Certainly national governments remain more important policymakers than the EU. Becoming a citizen of another EU country is, for example, still a national process. Still, what has occurred in a not even all that long a lifetime is nothing short of amazing. And it is hard to believe that all that change can simply unravel. But, then again, it was similarly inconceivable that Monnet's seeming pipe dream would have come as far as it has. Even Monnet would probably have been surprised. So, in the end, the future is almost always difficult and often impossible to predict.

Create Central

www.mhhe.com/createcentral

Additional Resources

Bulmer, Simon, and Lequesne, Christian, eds. (2013). *The Member States of the European Union* (Oxford University Press)

Explores how the EU has impacted its member countries individually.

McCormick, John, and Olsen, Jonathan. (2014). *The European Union: Politics and Policies* (Westview Press)

Piris, Jean-Claude. (2013). "The Five Crises in Europe and The Future of the EU," *E!Sharpe* online at http://esharp.eu/essay/28-the-five-crises-in-europe-and-the-future-of-the-eu/

Internet References . . .

Eurobarometer

Provides numerous polls that sample of opinion of people in the EU's member countries on a range of issues.

http://ec.europa.eu/public_opinion/index_en.htm

Europa

The official website of the European Union.

http://europa.eu/index_en.htm

E!Sharpe

An Internet site for news and opinions about the European Union.

http://esharp.eu/

United States Mission to the European Union

useu.usmission.gov/

Selected, Edited, and with Issue Framing Material by:
John T. Rourke, *University of Connecticut, Storrs*

ISSUE

Should the Export-Import Bank Be Eliminated?

YES: Veronique De Rugy, from "Examining Reauthorization of the Export-Import Bank: Corporate Necessity or Corporate Welfare?" Testimony During Hearings Before the Committee on Financial Services, U.S. House of Representatives (2014)

NO: Fred P. Hochberg, from "Examining Reauthorization of the Export-Import Bank: Corporate Necessity or Corporate Welfare?" Testimony During Hearings Before the Committee on Financial Services, U.S. House of Representatives (2014)

Learning Outcomes
After reading this issue, you will be able to:
• Explain why trade is important to the United States.
• Understand how trade deficits impact the overall balance of payments and, thereby, help determine a country's financial health.
• Discuss the role of the Export-Import Bank in fostering U.S. exports.
• Comment on whether or not the bank is doing a reasonable job in fulfilling its mission.

ISSUE SUMMARY

YES: Veronique De Rugy, a senior research fellow at the Mercatus Center at George Mason University, concludes that a close examination of the Export-Import Bank's activities and outcomes shows that it does not meet the standards of its own criteria, and the facts do not support these criteria for the continued activities of the Bank.

NO: Fred P. Hochberg, chairman and president of the Export-Import Bank of the United States, proclaims that he is proud of the work the bank's more than 400 employees do to empower U.S. companies and support American job growth, says the reforms are addressing some of the criticisms of the bank, and argues that the bank deserves to have its charter renewed for five years.

Foreign trade is ancient. Trading records extend back to almost 3000 B.C., and even before that archaeologists have uncovered evidence of trade in the New Stone Age, or Neolithic period (9000–8000 B.C.). Exports bring revenue into a country that stimulates its economy. Exports create and maintain jobs. Currently about 11 million American workers hold export-related jobs and make up nearly 7 percent of the U.S. workforce. Imports are also important for oil and other vital resources, for things we want, and to buy things inexpensively. For example, the cost of living for Americans would be considerably higher if inexpensive clothing, electronic goods, and many other items were not available through imports.

While foreign trade has always been important, it has increased in volume and become an even more significant factor in national prosperity and power in recent centuries, especially since World War II. World trade measured in exports totaled $70 billion in 1950. Since then, trade has grown rapidly, and in 2012, world trade stood at $22.6 trillion, some 32,300 percent more than in 1950. This $22.6 trillion included $18.3 trillion in "merchandise (or goods) trade," which involves tangible items ranging from agricultural products, to minerals such as petroleum, and manufactured products such as autos. "Services trade" accounted for another $4.3 trillion in 2012. Services include things that you do for others. When U.S. insurance companies earn premiums for insuring foreign assets or people, the revenue they generate constitutes the export of services. Among the world countries, the United States ($2.3 trillion) and China ($2.2 trillion) lead in the export of goods and services. China has a substantial lead in the export of merchandise over the United States, but the U.S. lead in services exports evened the overall figure.

Where the United States and China diverge is in imports. U.S. imports came to $2.7 trillion. This meant a U.S. trade deficit of about $500 billion in 2012. By contrast, China's imports were only $2.1 trillion, giving it a $100 billion trade surplus. One major cause of the U.S. trade deficit is the cost of petroleum, natural gas, and other energy imports. At over $400 billion in 2012, they equaled about 80 percent of the deficit. Another problem is the U.S. trade balance with China. U.S. imports of goods and services from China were almost four times the reciprocal importation of U.S. goods and service by China. This left the United States with a trade deficit of over $300 billion with China.

These numbers are important because these are a key factor in the U.S. "balance of payments," the total net flow of money in and out of a country for trade, investments, foreign aid, and other expenditures and revenue. Much like individuals or businesses, countries are financially advantaged if they have a positive balance of payments and disadvantaged by a negative (deficit) balance. Deficits are especially worrisome if they are chronic, as they are for the United States. The U.S. balance of payments deficit in 2012 was $115 billion. That is far better than the $201 billion deficit in 2006, but it remains a persistent deficit. Indeed, there has only been one year since 1982 that the U.S. balance of payments has not had an annual deficit. The major cause is the trade balance, which has been in chronic deficit since 1973.

The bottom line for this largely negative financial news with regard to the debate over the U.S. Export-Import Bank is that it is clearly in the U.S. national interest to do whatever it reasonably can to increase exports and decrease imports in the United States to reduce or eliminate the trade deficit as part of the effort to break even or even achieve a surplus in the balance of payments. On the import side, a major advance has been the decrease in the annual oil imports from a record high of 3.6 billion barrels in 2006 to 2.8 million barrels in 2013. At a recent price of $95 per barrel, that saves about $76 billion annually in imports. There are many efforts to also increase U.S. exports. The Export-Import Bank is part of the campaign. Veronique De Rugy in the YES selection of this issue gives a good review of the bank that need not be duplicated here. As a brief preview, though, the Export-Import Bank was created during the Great Depression years in 1934. Its role is to promote U.S. exports, and therefore American jobs, by financing or insuring foreign purchases of U.S. good where the potential foreign purchaser cannot secure the necessary financing through normal banking channels. The bank is a chartered corporation under U.S. law, and the hearings from which the selections herein are drawn were part of the process in 2014 by which Congress was considering whether to renew the bank's charter for another three years. As you will see in De Rugy's remarks, the bank has been criticized for favoring big over medium and small businesses, for poor accounting and other administrative processes, and for supporting some businesses, such as those involved in fossil fuel production, that seem at odds with the Obama administration's policy of curbing global warming (see Issue 15). Another charge is that guaranteeing loans given to foreign companies, even if it is to buy U.S. product, helps strengthen those foreign companies, many of which compete with U.S. companies. In the NO selection, Fred Hochberg, the bank's head since 2009, tells Congress that the bank is doing a needed job well, and also adds that it is making changes to meet some of the criticisms that have been voiced.

YES

Veronique De Rugy

Examining Reauthorization of the Export-Import Bank: Corporate Necessity or Corporate Welfare?

Introduction

In gathering to discuss the past and uncertain future of the US Export-Import Bank, we have a rare opportunity to reconsider the assumptions, mission, and activities of the federal government's official export credit corporation ahead of the reauthorization vote. In order to make the best decision, however, policymakers must know all the facts.

Before delving into the weeds of export credit finance, there are two fundamental realities to keep in mind.

First, export promotion programs, like Ex-Im finance, are not critical to US exports. Second, the Bank's mission is inherently contradictory.

First, Ex-Im yields a minuscule influence on US exports. At most, Ex-Im can claim to influence roughly 2 percent of both the value of total US exports and the total number of export-related jobs. Since the Bank's methodologies have been criticized by the General Accounting Office [GAO] and its own inspector general, the Bank's true influence is likely smaller.

Second, Ex-Im's mission refutes itself. The Bank's charter instructs administrators to extend assistance to projects that cannot find financing in private markets; these projects must also provide a reasonable chance of repayment. If a project cannot find private finance, it is probably too risky to repay the borrowed funds. But if a project has a good chance of repayment, it should easily find private finance. Any single project the Bank finances cannot meet both conditions of its charter.

This tension results in a bifurcation of the Bank's portfolio: profit-generating projects, like those involving Boeing, make up for losses in other areas. Ex-Im supporters likewise can either argue that (1) the Bank makes strong profits for the [U.S.] Treasury, or (2) the Bank provides needed but risky finance to important export opportunities. But if the Bank is making profits, that is an argument for privatization. If the Bank is suffering losses, that is an

argument for shutting it down. Neither scenario supports federal government involvement.

With these heavy caveats in place, I will now contribute to this conversation by providing contextual information about the Bank's history, processes, and portfolio. I will then consider the claims made by both Ex-Im supporters and critics. Specifically, I will examine the validity of arguments that the Export-Import Bank

1. plays a critical role in promoting US exports;
2. maintains or creates US jobs;
3. substantially benefits small business;
4. levels the playing field for US companies that compete against foreign companies that receive benefits from their countries' export credit agencies; and
5. is a good deal for taxpayers.

I conclude that a close examination of its activities and outcomes shows that the Export-Import Bank does not meet the standards of its own criteria, and the facts do not support these popular arguments for the continued activities of the Bank.

EX–IM Bank Basics

The Export-Import Bank has evolved since it was created in 1934 by President Franklin Delano Roosevelt. While its original purpose was to provide immediate trade financing with the Soviet Union, the Bank's focus shifted throughout the years, gradually changing from post-WWII reconstruction tool, to development bank for impoverished nations, to foreign policy tool, and eventually to export-promotion corporation. Ex-Im's mission today is to "aid in financing and to facilitate exports of goods and services, imports, and the exchange of commodities and services" between the United States and foreign countries and "in so doing, to contribute to the employment of United States workers" as described in the current charter that was last amended in 2012.

Veronique De Rugy. Testimony During Hearings Before the Committee on Financial Services, U.S. House of Representatives, June 25, 2014.

This 116-page document contains guidelines and proscriptions for the many pet programs that have developed in its 80-year history, including instructions regarding small business lending, green energy projects, engagement with sub-Saharan Africa, and prohibitions against aiding "Marxist-Leninist countries" or financing "defense articles."

The main tools provided to the Bank to achieve these ends are (1) loan guarantees, (2) working capital guarantees, (3) direct loans, and (4) export credit insurance.

Through loan guarantees, which presently constitute the largest portion of Ex-Im financing, the Ex-Im Bank assumes the majority of the repayment risk of the foreign buyer's debt obligations. The Working Capital Guarantee program guarantees short-term working capital loans made to qualified US exporters, through which the Ex-Im Bank assumes almost all the risk to lenders, which are usually commercial banks. The Ex-Im Bank's direct loan program provides loans directly to foreign buyers of US exports; if the foreign borrower defaults, the Ex-Im Bank will be responsible for the total value of the outstanding principal and interest on the loan. Finally, the Ex-Im Bank's export credit insurance program issues insurance policies to US exporters, often small businesses, which provide credit to the exporter's foreign buyer.

The Export-Import Bank's gradually expanding mission and authority produced a large but little-known federal corporation whose activities far exceed its original purpose.

Total Ex-Im authorizations, or the total amount of funding that the Bank commits to finance a higher total value of exports, increased from $12.37 billion in FY 2007 to $35.73 billion in FY 2012 before dropping a bit to $27.2 billion in FY 2013. [FY means fiscal year, the budget year. The U.S. FY 2014 extends from October 1, 2013, to September 31, 2014.] The Bank claims that this $27 billion in authorizations supported $37.4 billion worth of US exports worldwide during the same year. These amounts, however, significantly understate the financial risk that the Bank, and therefore taxpayers, are exposed to.

The total amount of exposure—defined by the Bank as "authorized outstanding and undisbursed principal balance of loans, guarantees, and insurance" plus "unrecovered balances of payments made on claims . . . under the export guarantee and insurance programs"—has grown consistently over time. Total Ex-Im Bank exposure grew from $57.42 billion in 2007 to $113.83 billion in 2013—without ever dropping, even when the corresponding level of authorizations dropped.

What benefits do US taxpayers receive from the Export-Import Bank's many activities? Do they outweigh their costs? I now turn to the five most popular claims made by supporters of the Ex-Im Bank.

Claim 1: The Export-Import Bank promotes exports

This claim has two branches: that the Ex-Im Bank improves the trade balance by filling an important "financing gap" in supporting US exports, and that it supports a significant share of exports. Neither has merit.

Some high-value projects go unfunded, the argument runs, because unusual or untenable investment risks scare off private financiers. The federal government should step in and fund the projects that the private market rejects. Without this federally provided export finance, US exports would be significantly dampened.

The assumptions behind this argument are inherently flawed because private investors are not likely to leave value on the table. It is proper that high-risk projects should not always find financing. Prohibitively high risk rates serve as a signal that investment funds could be more effectively spent elsewhere. Instead of making the difficult case that the Bank should subsidize losing projects, Ex-Im supporters prefer to use euphemistic language about "financing gaps" and "leveling the playing field." Nothing changes the fact, however, that these projects failed to attract private capital because their profit opportunities did not warrant investment.

Second, the data do not bear out the claims that the Export-Import Bank supports a substantial portion of US exports. Export data from the Census and the Export-Import Bank from 2000 to 2010 show that Ex-Im-backed activity accounts for approximately 2 percent of all US exports during that time. There are no grounds to claim that Ex-Im activity is critical to US exports.

It is has long been known that export credit corporations cannot substantially influence broader national trade outcomes. The GAO stated back in 1992 that "export promotion programs cannot produce a substantial change in the US trade balance, because a country's trade balance is largely determined by the underlying competitiveness of US industry and by the macroeconomic policies of the United States and its trading partners." [Therefore,] reforming the broader macroeconomic policies that are more likely to harm our trade position will help our exports far more than anything the Ex-Im Bank could do. The United States maintains the highest national statutory corporate tax rates among all [industrialized countries]. Reforming our punishing corporate tax rate would be one easy and feasible policy change that would pack a powerful punch without distorting markets and exposing taxpayers to risk, as the Export-Import Bank does.

Claim 2: The Export-Import Bank maintains and creates jobs

Supporters of federal programs often point to tangible outcomes, like the number of jobs created through federal spending, to justify their existence. Immediate employment effects are easily seen and therefore provide a potent shield against claims of ineffectiveness or waste. However, as the French economist Frédéric Bastiat first astutely pointed out over 150 years ago, good economists and students of public policy must also consider the *unseen* effects of government interventions to accurately perform a cost-benefit analysis. [all emphasis herein in the original]

Even if we take the Ex-Im Bank jobs claim at face value, the Bank influences a negligible number of export jobs. Since there were 11.3 million total export-related jobs in the United States in 2013, that means that the 205,000 jobs that Ex-Im claims to have created or supported represent only 1.8 percent of total export-related jobs. This is consistent with the small share of exports supported by Ex-Im.

Furthermore, this claim only represents the "seen" side of the story. To get a complete picture of the true economic effects of the Export-Import Bank's interventions, we must consider what would have happened if the Export-Import Bank did not spend money in this way. Would American citizens have spent their own funds in a more productive manner than their government did after taxing and distributing funds to the corporations that they chose? Might those decisions have created even more jobs than the ones for which the Export-Import Bank claims credit? There are a few reasons to think this could be the case.

First, foreign companies that receive Ex-Im financing are not necessarily purchasing *more* goods from US firms, but often simply buying different *kinds* of goods. Ex-Im interventions shift resources away from unsubsidized projects and toward artificially cheaper projects that the Bank subsidizes. Many of the jobs the Bank claims to "create" are in reality redirected from unsubsidized firms. The Export-Import Bank disadvantages employees of unsubsidized companies for the benefit of employees of subsidized companies.

These unfortunate consequences have long been known to Washington. A Congressional Budget Office (CBO) report from 1981 explains that under normal economic conditions "subsidized loans to exporters will increase employment in export industries, but this increase will occur at the expense of non-subsidized industries: the subsidy to one industry appears on other industries' books as increased costs and decreased profits."

As we will see later, despite the talk about "financing gaps," most of the loan value backed by the Bank benefits large and well-established companies that have ample alternative financing options. Even small businesses that receive support were often profitable well before Ex-Im operatives came knocking.

Washington has long known about the Bank's tendency to favor familiar firms. President [Ronald] Reagan's former Office of Management and Budget (OMB) director, David Stockman, had more than a few run-ins with the Export-Import Bank and its well-connected corporate beneficiaries as he tried to retire or reform the Bank in the early 1980s. In his words, the idea that export subsidies will create jobs or increase GDP is yet another example of "the single-entry-bookkeeping mentality that has larded the federal budget" with so many subsidies and payments to special interests. Despite its economic illiteracy, it is, for some, a powerful argument to maintain the status quo.

Last but not least, the federal government's own GAO finds that the Bank's job calculation method leaves much to be desired. A May 2013 report criticized the Bank for concealing the many methodological weaknesses that underlie its attractive "205,000 jobs" number. The GAO reports that Ex-Im extrapolates its numbers from the Bureau of Labor Statistic data product that measures how much of the output (revenue) of an industry goes into the input (production) of another industry and its employment equivalent. There are many limitations to this methodology such as the fact that it does not distinguish between full-time, part-time, and seasonal employment, thus painting a rosier, but inaccurate, picture.

Claim 3: The Export-Import Bank supports small businesses

The Export-Import Bank frequently boasts of its outreach and support to small businesses, but these claims are very misleading. The Bank selectively emphasizes portfolio numbers to overstate small business activities, employs an extraordinarily expansive definition of the term "small business," and disperses the majority of its largesse to large, established corporations.

Ex-Im's FY 2013 Annual Report states that "the Bank approved a record 3,413 transactions, or 89 percent, for small businesses." This statement is curiously incomplete. It is true that 89 percent of the total number of deals involved firms that fit the Bank's definition of a "small business." However, when you look at the total *amounts* that the Bank distributes, the picture is decidedly different. During FY 2013, for instance, only $5.2 billion of the $27.3 billion in total authorization amounts, or

19 percent, was designated as "small business" activity on the Ex-Im Bank's annual report. This is concerning, as the Bank's own charter mandates that no less than 20 percent of total authority for each fiscal year be directly made available to small business concerns.

Additionally, the Bank employs a rather unconventional definition of "small business." While the term invokes images of mom-and-pop stores and enterprising startups, Ex-Im's definition is considerably more expansive than that of the Small Business Administration (SBA). With few exceptions, the SBA sets its "small business" threshold at firms with fewer than 500 employees and no more than $7 million in average annual receipts—criteria much stricter than the Bank's. Ex-Im's definition of "small" includes manufacturing and wholesale firms that employ anywhere from 500 to 1,500 workers, general construction firms that earn anywhere between $13.5 million and $17 million a year, and in other sectors, firms with annual revenues of up to $21.5 million.

According to the Bank's data for FY 2013, much of its direct and indirect subsidies benefit giant manufacturers and well-connected exporters. For instance, America's number one exporter, the Boeing Corporation, received roughly 66 percent of the value of all loan guarantees last year. The Bank reports that Boeing was designated primary exporter for 55 deals, valued at roughly $8.3 billion in total assistance. This means that upwards of 30 percent of the value of all Export-Import Bank assistance in 2013 directly benefited the Boeing Corporation. Additionally, the top 10 exporter beneficiaries of Ex-Im assistance—among whom we find General Electric, Dow Chemical, Bechtel, Caterpillar, and John Deere—received a combined total of roughly 75 percent of the value of total assistance last year.

Claim 4: The Export-Import Bank levels the playing field for US exporters competing with foreign corporations subsidized by their own governments

Another popular argument is that the Export-Import Bank is necessary to counteract the competitive disadvantages posed by the export credit agencies of foreign nations. Since foreign firms enjoy the benefits of their own national export subsidy organizations, the argument goes, US firms would struggle to compete internationally if our federal government did not provide similar aid. The US Export-Import Bank is a necessary, although perhaps unfortunate, weapon in this international export subsidy arms race.

Few would disagree that it would be much preferable if US exporters only had to compete on price and quality against unsubsidized foreign companies. However, the fact that other countries choose to engage in bad economic

policy does not immediately justify the United States following suit. This argument might hold water if (1) a substantial portion of the Export-Import Bank's portfolio is indeed dedicated to counteracting the effect of foreign export credit agencies; (2) a substantial portion of total US exports depend on Ex-Im assistance to successfully compete on a global scale; and (3) engaging in suboptimal policies because other countries do so makes economic sense.

If the Export-Import Bank were truly a critical tool in counteracting the activities of foreign export credit agencies, the bulk of the Bank's portfolio should to be dedicated to this purpose—which is not the case.

For a long time, the Bank did not specify the exact purpose of each transaction in its reports, so it was difficult to determine how much of its portfolio was dedicated to which goals. However, Congress recently obliged the Bank to provide more explanations for certain portfolio transactions. According to the Bank's 2013 Annual Report, only $12.2 billion of the $37.4 billion portfolio—that is, less than one-third—was dedicated to counteracting competitive disadvantages wrought by foreign export credit agencies. And, as previously mentioned, if export subsidies were truly so critical to competing abroad, then we would expect far more than 2 percent of total US exports to receive Ex-Im assistance. Yet somehow, the other 98 percent of unassisted US exports thrive and successfully compete in the global marketplace.

Ironically, subsidies also damage recipients in the long run by dulling their competitive edge. Subsidized businesses often grow complacent and lazy because they know they can rely on government assistance. When markets change and pressures mount, subsidized industries find that they simply cannot keep up—so they come to Congress for a bailout.

Most importantly, subsidizing exports actually harms US consumers and helps foreign consumers. Indeed, this is one of the selling points of Ex-Im programs. When foreign export agencies subsidize their exporters, they actually help US consumers at the expense of their own citizens. In subsidizing our own exports in response to foreign subsidies, the Export-Import Bank actually hurts US consumers more.

Claim 5: The Export-Import Bank is a good deal for taxpayers

Defenders of the Export-Import Bank often point out that the agency does not cost anything to taxpayers; in fact, they argue, the Bank actually generates profits for the US Treasury. In FY 2013, for example, Ex-Im claimed to return over $1 billion in profits to the Treasury. Regardless of whether the federal government should or must operate an export-credit agency, the fact that the Export-Import Bank reports profits to the government is enough of a reason for many to support it.

[However,] even if the Bank were profitable today, we could not rely on the Bank to earn profits for American taxpayers for years to come. The Bank ran a deficit several years in a row in the 1980s and was reported to have considered requesting a $3 billion bailout in 1987. This could happen again. [Also,] financial projections are only as sound as the methodologies employed to produce them. Many have been skeptical of the Bank's risk analyses, default assumptions, and accounting methods for years. Some of the most vociferous criticisms have come from within the Export-Import Bank [inspector general], as well as from the GAO and CBO. The errors that these criticisms reveal undercut Ex-Im supporters' claims of profitability and sustainability and in fact reveal that the Bank exposes taxpayers to the risk of "severe portfolio losses," in the words of the Bank's own inspector general.

An alarming 2012 report from the Inspector General of the Export-Import Bank reveals that the Bank does not employ adequate risk analyses. The Bank's unique position as a governmental export credit insurer and underwriter exposes it to a number of unusual risks. In addition to balancing normal market considerations and operational boundaries imposed through legislation, the Bank must also adequately calculate and mitigate taxpayer exposure to credit risks, political risks, currency risks, and various concentration risks riddling its vast portfolio. The inspector general warns that "Ex-Im lacks a systemic approach to identify, measure, price, and reward" these many portfolio risks. This improper loss reserve methodology may have "resulted in the systematic under-reserving and underpricing of the portfolio risk," which significantly limits the predictive veracity of the Bank's projections.

Improper accounting may yield impressive illusions in the short term but will not prevent financial disaster in the event of an unexpected downturn. For many years, Ex-Im skeptics could only point to the dire pleadings of the GAO and Inspector General to bolster their case that the Bank's "profits" may be a misleading accounting fiction. The collapse of the US housing market, prompted partially by the risky underwriting of federal mortgage corporations Freddie Mac and Fannie Mae, provides another cautionary tale.

We do not need to wait for an economic collapse to reveal the financial fallout of the Export-Import Bank's phony accounting. A new CBO report from May of 2014 confirmed the GAO's worst suspicions. The CBO found that the Export-Import Bank's promised budget savings of over $14 billion over the next decade are illusory. The CBO's more accurate accounting method shows that that these Export-Import Bank programs will actually *cost* taxpayers $2 billion over the next decade.

The bottom line is that supporters of the Export-Import Bank can no longer truthfully claim that the Bank's programs are guaranteed to yield a profit to the federal coffers.

Conclusion

The data provide no support for the most popular arguments for the Export-Import Bank.

The Export-Import Bank cannot be shown to "maintain or create" any number of US jobs because the Bank's job calculation methodology is so weak that the GAO had to provide an in-depth audit to shake the Bank into reform. While the Bank at least now includes a brief exposition of these weaknesses deep in its Annual Reports, it insists on continuing to use the old, flawed methodology. Even this flawed job measure constitutes only around 2 percent of the total number of export-related jobs each year. What's more, for every job the Bank can claim to have "created," there are numerous employees of unsubsidized competing firms who can claim to have been hurt by this action.

The Bank cannot claim to play a "critical role in promoting US exports" as the total export value of its annual portfolio only accounts for roughly 2 percent of the total value of US exports each year. The Bank cannot claim to be filling the purported "financing gap," either, as most of its portfolio assists large, connected firms that would have no trouble procuring alternative financing. Finally, economists have long known that export credit agencies have, at best, a tiny effect on national exports compared to potent broader economic trends.

The Bank does not substantially benefit small businesses but disproportionately benefits large corporations like Boeing and General Electric. Supporters like to point to the number of deals given out to small businesses, but in terms of dollar amounts, huge multinational corporations are the clear winners. Even the "small businesses" that Ex-Im claims to support are not exactly "small": Ex-Im's definition of "small business" is far more expansive than the one employed by most federal offices and can include businesses that receive up to $21.5 million in revenues each year.

The Bank cannot claim to level the playing field for US companies that compete against foreign subsidized competition, either. For one, the United States is the number two export credit subsidizer in the world, beaten only by China. The Bank's own records report that only 30 percent of the estimated export value in its total portfolio goes toward this kind of activity. Even if the Bank's entire portfolio were turned to this purpose, this position is economically untenable. By subsidizing exports, the Export-Import Bank helps foreign consumers and hurts American consumers, taxpayers, and borrowers.

The Bank cannot point to profits for taxpayers because its internal reporting is significantly flawed and CBO analyses show that the Bank will actually cost taxpayers billions of dollars over the next decade. What's more, numerous audits by the CBO, the GAO, and the Bank's own Inspector General show that the Export-Import Bank's internal risk management and accounting practices are woefully inadequate and leave taxpayers exposed to massive liabilities.

The Export-Import Bank fails on each count. An entity that neglects to meet even its supporters' own criteria simply has no justification for existence and no claim to reauthorization. The Bank has long outlived its purpose and cannot manage to meet the standards of the new missions that have been developed to validate its existence. For policymakers who have the facts, the choice is clear: the Export-Import Bank must go.

Veronique De Rugy is a senior research fellow at the Mercatus Center at George Mason University. She holds a PhD in economics from the Pantheon-Sorbonne University in France.

Fred P. Hochberg  **NO**

Examining Reauthorization of the Export-Import Bank: Corporate Necessity or Corporate Welfare?

[I am here] to testify before you as the Committee considers the progress that the Export-Import Bank of the United States ("Ex-Im Bank" or "the Bank") has made in supporting U.S. jobs through exports since our last reauthorization just two short years ago.

The Administration is requesting a five-year reauthorization of the Export-Import Bank of the United States and a cap of $160 billion. This proposal reflects a stepped increase totaling $20 billion above the [current] exposure cap of $140 billion. The proposal reflects a 2.9 percent annual increase over [this year] and a 14 percent increase over all 5 years. The Bank's mission is to support jobs in the United States by facilitating the export of U.S. goods and services. The reauthorization of the Bank at the $160 billion cap supports an estimated 1.3 million U.S. jobs between 2015 and 2019. The Bank has identified four main macroeconomic factors (competition from other Export Credit Agencies (ECAs), financing for small business exporters, financing for developing/emerging markets, and increasing regulations) that show the continued need for Ex-Im Bank financing support for U.S. exports. Ex-Im evaluated several scenarios, and determined that the $160 billion requested level will provide the needed capacity to continue supporting small business, and fill gaps in trade-financing.

I am proud of the work our 400+ employees do each day to empower U.S. companies and support American job growth. Since our last reauthorization in 2012, Ex-Im Bank has supported nearly half a million American export-backed jobs, while generating nearly two billion dollars for the taxpayers and maintaining a low default rate [on loans] of .211 percent. For perspective, our default rates over the last five years, during the height of the financial crisis shown below illustrate a very responsible approach to our portfolio.

I would like to focus today on three primary issues.

- Requirements set forth in our last reauthorization, including our enhanced risk posture.

- Our impact on small business.
- The heightened need for a robust Ex-Im Bank given the brutally competitive global trade environment faced by American businesses.

Reforms Set Forth in Our Last Reauthorization, Including Our Enhanced Risk Posture

Since our last reauthorization, Ex-Im Bank has met all of the reporting requirements set forth in our reauthorization bill. We produced numerous reports for this Committee.

And the Bank implemented additional reforms aimed at ease of service and minimizing any negative effects on U.S. industry including:

- Increased access to co-financing to better support globalized supply chains
- Streamlined content procedures by providing for annual certifications
- Revised economic impact procedures to airline services

Ex-Im Bank fills gaps in private sector financing and only gets involved when the private sector is unable or unwilling to provide financing or when there is foreign ECA competition. Last year, Ex-Im Bank provided $27.3 billion in authorizations, down from the previous year. This supported $37.4 billion in export value.

We have worked well with the Government Accountability Office (GAO) on several reports that they were tasked with, including reviews of the Bank's Business Plan, [its] jobs calculation methodology, [and its] risk management.

I met with [the head of the GAO] Comptroller General Gene Dodaro to indicate my personal commitment to working cooperatively and in an open and transparent fashion with the GAO on audits. The Comptroller General

Fred P. Hochberg. Testimony During Hearings Before the Committee on Financial Services, U.S. House of Representatives, June 25, 2014.

thanked me for arranging the meeting and noted that very few agency heads take the time to come meet with him. He also added that his staff indicated they have a good working relationship with Ex-Im Bank staff.

We have also made a number of changes as a result of recommendations from our Inspector General (IG) and the GAO reviews [to ensure that the loans that the Bank makes are not overly risky and to improve measuring and reporting to Congress related to the cost and impact of the Bank's activities]. [Among other steps,] we created the position of, and hired, a Chief Risk Officer (CRO) who reports directly to me as the President and Chairman. We [also] established a risk committee called the Enterprise Risk Committee (ERC) that assesses comprehensive risk issues and reports semi-annually to the Bank's Board of Directors, while providing the other Directors and me with monthly updates.

Our Positive Impact on Small Business

As a small businessman for 20 years and as a former Acting Administrator of the Small Business Administration, I place a premium on the role that small businesses play in our economy. One of the core missions of Ex-Im Bank is ensuring that small businesses are at the forefront of U.S. exports. We cannot grow our economy without fully supporting the small businesses of America in competing globally.

In 2013, the Bank financed a record 3,413 small businesses—nearly 90 percent of Ex-Im's transactions. This amounted to $5.2 billion in direct small business financing. When you examine the export value of small business exports as a percentage of our overall portfolio export value, of the $37 billion of exports supported by Ex-Im Bank, nearly 33 percent were small businesses.

In addition, Ex-Im financed more small businesses over the last five years than in the prior eight years combined. The Bank also financed more minority- and woman-owned businesses in the last five years than over the prior sixteen years combined.

The Bank also supports tens of thousands of small businesses whose goods are incorporated into larger export products such as transportation, heavy machinery, oil and gas facilities and other manufactured goods. Industry frequently refers to these small business exporters as "hidden exporters." It is important to note that the manufacturing process in the United States has changed significantly over the last few decades. Many more companies use vast supply chains to produce finished goods, as evidenced by our largest exporters. When you see the name of a large company on a name plate of finished goods, there are thousands of small businesses that contributed to that finished product.

Ex-Im Bank supports American jobs in countless cities and towns across America. Since 2009, Ex-Im Bank supported $20 billion in authorizations for Texas businesses and no city in the nation exported more Ex-Im Bank financed goods and services than Houston. We also financed $9 billion in authorizations for California businesses, $1 billion in authorizations for South Carolina businesses, and approximately $500 million in authorizations for Missouri businesses over the past five years—just to name a few. Ex-Im Bank supports exports in all 50 States and Territories.

The Need for Ex-Im Bank Given the Brutally Competitive Global Trade Environment

There are some 60 Export Credit Agencies (ECAs) around the world. Every industrialized country has its own version of Ex-Im, and each is tasked with supporting the domestic exports of their respective nations.

The mission of Ex-Im Bank is to empower U.S. companies—large and small—to turn export opportunities into sales that maintain and create U.S. jobs and contribute to a stronger national economy. When private sector funding is unavailable or there is foreign ECA involvement, Ex-Im Bank provides export financing through its loan, guarantee, and insurance programs in order to level the playing field for U.S. businesses.

Let me provide a few examples.

On a recent visit to Chicago, I met with Mary Howe, a fourth generation small business owner and owner of Howe Corporation. We provide trade credit insurance to Howe Corporation so they can export their ice making machines around the globe. Trade credit insurance enables U.S. exporters, most of them small businesses, to compete and sell overseas with their receivables insured for loss. We provide insurance when the private sector does not, because frequently the dollar volume of small businesses is too small to be of interest to private insurers. Credit insurance is part of running a responsible business, just like fire or theft insurance.

Another example is the working capital guarantee Ex-Im provides to Auburn Leather in Auburn, Kentucky, a town with a population of 1,332 people as of 2012. Auburn Leather makes the laces that are in baseball gloves and boat shoes. Over the years, I've come to know the owner, Lisa Howlett and I was delighted that Ex-Im could provide a working capital guarantee so that her business could have the liquidity to meet large foreign purchase orders. Ex-Im exists to fill gaps in private sector financing. Nowhere else does this show itself than in insurance for small businesses. These small businesses often grow out of needing Ex-Im Bank and will later obtain private sector financing—which we encourage.

In this global economy, buyers will make procurement decisions based on the availability and attractiveness of financing. Ex-Im Bank provides long-term direct loans and guarantees to foreign buyers of capital goods, so that their best option is to buy a product with the "Made in America" label on it. Our financing supports the export of products from U.S. companies such GE Transportation, so that the sales and jobs arising from their locomotives will benefit workers in Erie and Grove City, Pennsylvania, rather than in Beijing, China.

Ex-Im Bank also provides support, if necessary, to level the playing field when financing is provided by foreign governments to their companies who compete against U.S. exporters. Critics of the Bank argue that Ex-Im should not be involved in helping to level the playing field for U.S. businesses and that the private sector should handle this. I concur. When the private sector can provide financing, we prefer that course of action. Ex-Im Bank does not compete with private sector lenders, but rather provides financing for transactions that would otherwise not take place because commercial lenders are either unable or unwilling to provide financing support. Often, private lenders do not want to get involved with small business exports because the dollar value is too low. Another area is structured finance, which generally requires more than twelve years of financing.

The last five years have seen major shifts in the global financing landscape that could have a serious impact on American competitiveness and economic growth in the years ahead. But despite the global recovery, commercial banks still have not fully regained their appetites for offering the kind of long-term financing necessary to fuel many export projects. As banks withdraw from important areas of export finance just as exports become more vital to economic growth worldwide, ECAs are rising in importance for many nations. More and more, export financing is occurring outside of the international standards put in place by the OECD [Organization of Economic Cooperation and Development] to avoid a financing "race to the bottom." The OECD Arrangement on Officially Supported Export Credits, while not perfect, constitutes the international export credit framework that sets the most generous terms and conditions governments may provide when financing exports. Countries like Russia and China that are not members of the OECD and do not operate within the international export credit framework may offer financing terms that distort the global market. The effect of this can be an increasingly unleveled playing field for American exporters. When the private sector is unable or unwilling to provide financing support, Ex-Im Bank steps in to level the playing field so that American businesses can win their fair share of export orders.

Conclusion

I want to thank this committee for their work on our charter in 2012, and stress the importance of a timely reauthorization in 2014. As I mentioned earlier, there are some 60 Export Credit Agencies around the globe. Make no mistake, these foreign governments want for themselves the 205,000 American jobs Ex-Im financing helped support last year.

As I travel the world on behalf of American companies, I know that my counterparts in China, Brazil, Russia, South Korea, and many others are right behind me, fighting for business. These nations, and many others, are serious competitors in the global marketplace—and each supports its exporters aggressively.

For example, the South Korean ECA, supporting an economy less than one tenth of our size, finances more than 3 times the exports for South Korean companies than the Ex-Im Bank finances for U.S. companies. A few weeks ago, we heard from one exporter at a Capitol Hill roundtable whose business was adversely impacted by the actions of a foreign ECA. Steve Wilburn is a veteran whose small renewable energy company, FirmGreen, has created 165 jobs in California and at suppliers in seven other states since partnering with Ex-Im. A $48 million loan from Ex-Im empowered FirmGreen to bring their innovative ideas, equipment, and services to Brazil. And the Novo Gramacho biogas project will convert dirty methane gas into clean, compressed biomethane gas. That is not just good for the environment. It is good for the U.S. economy, and the communities across America made stronger thanks to an infusion of new jobs. I had the opportunity to visit the Novo Gramacho site not long ago—this is a game-changing technology. Steve is looking at two more sites in Brazil. His company has emerged as a global leader in its field.

But two months ago, Steve was stunned to hear that he lost out on a $57 million project in the Philippines. He had been told he was the preferred supplier. He thought he had it in the bag. But he lost out because his competitor from Korea convinced the buyer that Steve's business might not have the financing to get it done. They pointed to the debate surrounding Ex-Im's reauthorization, and they said: "There's too much uncertainty there—you can't rely on America." Can't rely on America? That's just wrong.

There is a strong drive to increase exports from many countries around the globe. We need to send the same signal to competitor nations that we stand behind American workers and exporters—and ensure that products stamped "made in the U.S.A." are able to compete on a level playing field. In order for U.S. businesses to be able to compete based on the price and quality of their exports, Ex-Im needs to be there to level the playing field when it comes to meeting foreign ECA competition.

The thousands of businesses that benefit from Ex-Im Bank financing—almost 90 percent of which are small businesses—appreciate the fact that Congress was able to reach an agreement to reauthorize the Bank in 2012. They need to know that we will be around in the years ahead to support them as they face off against increasingly intense foreign competition, grow their exports, and create more jobs here at home.

Every other country is strengthening the capacity of their ECAs to support their domestic exporters. Only the United States is having a conversation about actually making Ex-Im Bank less robust in response to global trends. This uncertainty is only emboldening foreign ECAs in their effort to take jobs away from U.S. businesses.

FRED P. HOCHBERG is chairman and president of the Export-Import Bank of the United States. He has an MBA degree from Harvard University.

EXPLORING THE ISSUE

Should the Export-Import Bank Be Eliminated?

Critical Thinking and Reflection

1. Should the government be in the business of backing loans that are so risky that companies and private banks will not capitalize them?
2. If the Export-Import Bank is making loads that have, as it says, a default rate as low as 2 percent, when the average default rate on regular commercial loans is about 3 percent, is the bank actually making loans that other banks would not?
3. Does it make any difference whether the Export-Import Bank mostly supports loans to foreign buyers of goods from large U.S. companies as long as the purchases make or protect American jobs?

Is There Common Ground?

Somewhat unexpectedly the question of renewing the charter of the Export-Import Bank became a hot topic in 2014. Part of the cause was the fault found with the bank's procedures by the General Accountability Office. Part was the charge that the bank was favoring large corporations over smaller ones. Partisan politics also heated the issue up, with Democrats defending the bank's operation and the Obama administration's recommendation for a five-year charter renewal, and Republicans castigating the bank and pondering only a short-term renewal, perhaps even less than a year, to force the bank to institute reforms. Prospects for a long-term renewal dimmed when House Financial Services Committee Chairman Jeb Hensarling, who led the hearings that were the forum for the selections above, strongly opposed renewing the Export-Import Bank. He called for at least fundamental overhaul of the bank and labeled it as "an offensive form of corporate welfare."

As of this writing in September 2014 with Congress nearing adjournment and with the Export-Import Bank charter about to expire, House Republicans have inserted a clause in the spending bill that will allow the bank to continue to operate until June 30, 2015. That signals some willingness of the Republican leadership to eventually reauthorize the bank as long as it makes the changes they wish. It seems likely this measure will pass and that 2015 will feature round 2 in the debate over the bank's future.

Create Central

www.mhhe.com/createcentral

Additional Resources

Bandow, Doug. (May 5, 2014). "Close the Export-Import Bank: Cut Federal Liabilities, Kill Corporate Welfare, Promote Free Trade." *Forbes*.

Hufbauer, Gary Clyde, Fickling, Meera, and Wong, Woan Foong. (May 2011). "Revitalizing the Export-Import Bank." Policy Brief PB11-6, Petersen Institute for International Economics, online at http://wwww.iie.com/publications/pb/pb11-06.pdf

Ikenson, Daniel J. (2014). "Export-Import Bank Debate Suggests Traditional Opponents of Corporate Welfare See Tea Party as Bigger Evil." CATO Insitute website at: http://www.cato.org/publications/commentary/export-import-bank-debate

This article highlights the irony that congressional liberals who once attacked the bank now support it, while some Republicans have turned against it under pressure from the Tea Party wing of the party. The article first appeared in *Forbes* on July 3, 2014.

Ilias Akhtar, Shayerah, Carpenter, David H., Levit, Mindy R., and Taylor, Julia. (August 1, 2014). *Export-Import Bank Reauthorization: Frequently Asked Questions*. Congressional Research Service, Report 443671, online at: http://fas.org/sgp/crs/misc/R43671.pdf

Internet References . . .

Export-Import Bank of the United States

On the website of the National Association of Manufacturers

http://www.nam.org/Issues/Trade/Ex-Im-Bank.aspx

www.exim.gov/

Tweets on EximBankUS

https://twitter.com/EximBankUS

Export-Import Bank: Status of GAO Recommendations on Risk Management, Exposure Forecasting, and Workload Issues

http://www.gao.gov/products/GAO-14-708T

Unit 4

UNIT

Armaments, War, and Terrorism Issues

*G*lobal peace has for most of human history been an idea that has been primarily a subject for philosophers, religious leaders, and other thinkers rather than political leaders. That began to change a little more than a century ago when rapid advances in weapons technology enabled warring peoples to kill one another in ever larger numbers. By the mid-1800s the first attempts were underway to create treaties to prevent wars if possible and limit them if not. For example, the Geneva Convention of 1864 was the first treaty to detail how wounded prisoners of war should be treated. Then the Hague Convention of 1897 created a Permanent Court of Arbitration to settle international disputes and, if wars did occur, it contained such restraints on their conduct as barring shooting from or dropping explosives from balloons.

Of course, wars have not ceased and the power of weaponry has grown exponentially. The nuclear age brought humans the literal capability of wiping themselves out. Responding to these changes, the efforts to prevent and limit wars and other form of violence have also grown greatly. Now the process is ongoing. Numerous arms control treaties have been negotiated and gone into effect. Other such treaties, such as the Comprehensive (Nuclear) Test Ban Treaty (CTBT) await the ratification of a handful of countries in order to become international law. There have also been additional treaties about the treatment of prisoners of war. New courts to settle disputes have been created and new regional organizations, such as the Organization of American States, and global organizations, such as the United Nations, have been founded primarily to keep peace.

It is possible to debate how much impact these efforts to prevent or regulate war have had, but it is clear that they have not achieved their ultimate goal of peace. The CTBT remains an open issue. New weapons technology, such as drone warfare, have arisen and need to be regulated. Some types of violence, such as terrorism, have become more prominent, and it and such questions as who is a terrorist and who is not need to be addressed. This reality of work to be done and questions to be resolved is the focus of this unit.

Selected, Edited, and with Issue Framing Material by:
John T. Rourke, *University of Connecticut, Storrs*

ISSUE

Should the United States Ratify the Comprehensive Nuclear Test Ban Treaty?

YES: Ellen Tauscher, from "The Case for the Comprehensive Nuclear Test Ban Treaty," Remarks at the Arms Control Association Annual Meeting at the Carnegie Endowment for International Peace, U.S. Department of State (2011)

NO: Baker Spring, from "U.S. Should Reject Ratification of the Comprehensive Test Ban Treaty," *The Heritage Foundation Web Memo #3272 (2011)*

Learning Outcomes
After reading this issue, you will be able to: • Relate the background of the debate over whether to ratify the Comprehensive Nuclear Test Ban Treaty (CTBT).   • Explain why the Obama administration wants the Senate to ratify the CTBT.   • Lay out the arguments against ratifying the treaty.   • Detail the procedural and political barriers to the Senate ratifying the treaty.

ISSUE SUMMARY

YES: U.S. Under Secretary of State for Arms Control and International Security Ellen Tauscher expresses the view that the United States will lose nothing and gains much by ratifying the Comprehensive Test Ban Treaty.

NO: Baker Spring, the F. M. Kirby Research Fellow in National Security Policy at The Heritage Foundation, asserts that the problems with the Comprehensive Test Ban Treaty that led the U.S. Senate to reject it in 1999 have, if anything, worsened in the intervening years.

A blinding flash, doomsday roar, and destructive pressure wave announced the first atomic weapons blast, the U.S. test near Alamogordo, New Mexico, on July 16, 1945. Following this first atomic test, the annual number of tests mushroomed to 178 in 1962. Then testing began to ebb in response to a number of arms control treaties beginning with the 1963 U.S.–U.S.S.R. Limited Test Ban Treaty prohibiting nuclear weapons tests in the atmosphere, in outer space, or under water; a declining need to test; and increasing international condemnation of those tests that did occur. The number of annual tests dipped into single numbers in 1992 for the first time in 35 years, and since 1996 there have been no American, British, Chinese, French, or Russian tests. Then in 1998 a series of tests marked India and Pakistan's acquisition of nuclear weapons. The most recent tests as of late 2014 were conducted in 2006, 2009, and 2013 by North Korea. With them, North Korea became the eighth country with an acknowledged nuclear arsenal (Israel has an unacknowledged arsenal). Because of the frequent secrecy of tests, there are disputes about the exact number of tests that have occurred, but the reputable Stockholm International Peace Research Institute (SIPRI) puts the total at 2,054.

One of the major efforts in arms control has been to ensure that no further nuclear weapons tests will ever occur. The centerpiece of that goal is the Comprehensive Test Ban Treaty (CTBT), which bans all such tests. The treaty was concluded in 1996, and 155 countries (79 percent of all countries) have ratified it as of early 2012. Nevertheless, the CTBT remains in limbo because it will not become operational until all the 44 countries that had nuclear reactors in 1996 ratify it. Several such countries, including the United States and China, have not ratified it. With President Bill Clinton in the White House, the United States signed the treaty in 1996, but the Senate rejected it in 1999. The vote of 48 yeas to 51 nays fell far short of the two-thirds vote necessary to ratify a treaty.

The Senate action has not meant new U.S. tests, however. The last of these 1,032 tests conducted occurred in 1992, and in the first selection U.S. Under Secretary of State for Arms Control and International Security Ellen Tauscher argues that the United States no longer needs to conduct nuclear explosive tests and, therefore, the Senate should reconsider its 1999 vote and ratify the CTBT. In the second selection, Baker Spring, a national security expert at the Heritage Foundation, disagrees. He contends that the reasons that the Senate rejected the CTBT remain relevant and that the effort to revive the treaty is an attack on the Senate's integrity.

YES ⬅

Ellen Tauscher

The Case for the Comprehensive Nuclear Test Ban Treaty

Many of you have heard me speak many times about what this Administration intended to accomplish and what we have accomplished. In the two years since President [Barack] Obama's speech in Prague, the Administration has taken significant steps and dedicated unprecedented financial, political, and technical resources to prevent proliferation, live up to our commitments, and to move toward a world without nuclear weapons. [In an April 2009 speech in Prague, the Czech Republic, President Obama said, "To put an end to Cold War thinking, we will reduce the role of nuclear weapons in our national security strategy and urge others to do the same."]

Under the President's leadership, we have achieved the entry into force of the New START agreement, adopted a Nuclear Posture Review that promotes nonproliferation and reduces the role of nuclear weapons in our national security policy, and helped to achieve a consensus Action Plan at the 2010 Nuclear Nonproliferation Treaty Review Conference.

The Administration also convened the successful 2010 Nuclear Security Summit, helped secure and relocate vulnerable nuclear materials, led efforts to establish an international nuclear fuel bank, and increased effective multilateral sanctions against both Iran and North Korea. As for what's next, our goal is to move our relationship with Russia from one based on Mutually Assured Destruction to one on Mutually Assured Stability. We want Russia inside the missile defense tent so that it understands that missile defense is not about undermining Russia's deterrent.

Even though this is a bipartisan goal—President [Ronald] Reagan and President [George H. W.] Bush both supported missile defense cooperation—it will not be easy. I know that many of you have opposed missile defenses. I have as well when the plans were not technically sound or the mission was wrong. But this Administration is seeking to turn what has been an irritant to U.S.–Russian relations into a shared interest. Cooperation between our militaries, scientists, diplomats, and engineers will be more enduring

and build greater confidence than any type of assurances. We are also preparing for the next steps in nuclear arms reductions, including—as the President has directed—reductions in strategic, non-strategic, and non-deployed weapons. We are fully engaged with our allies in this process.

But let me turn to the Comprehensive Test Ban Treaty. President Obama vowed to pursue ratification and entry into force of the CTBT in his speech in Prague. In so doing the United States is once again taking a leading role in supporting a test ban treaty just as it had when discussions first began more than 50 years ago.

As you know, in the aftermath of the Cuban Missile Crisis, the United States ratified the Limited Test Ban Treaty, which banned all nuclear tests except those conducted underground. The Cuban Missile Crisis, which was about as close as the world has ever come to a nuclear exchange, highlighted the instability of the arms race. Even though scholars have concluded that the United States acted rationally, the Soviet Union acted rationally, and even [Cuba's President] Fidel Castro acted rationally, we came perilously close to nuclear war. Luck certainly played a role in helping us avoid nuclear catastrophe.

In the months after the crisis, President [John F.] Kennedy used his new found political capital and his political skill to persuade the military and the Senate to support a test ban treaty in the hopes of curbing a dangerous arms race. He achieved a Limited Test Ban Treaty, but aspired to do more. Yet, today, with more than 40 years of experience, wisdom, and knowledge about global nuclear dangers, a legally binding ban on all nuclear explosive testing still eludes us. This being Washington, everything is seen through a political lens. So before discussing the merits of the Treaty, let me talk about this in a political sense for a moment. I know that the conventional wisdom is that the ratification of New START [Strategic Arms Reduction Treaty] has delayed or pushed aside consideration of the CTBT. I take the opposite view.

The New START debate, in many ways, opened the door for the CTBT. Months of hearings and debate and

Tauscher, Ellen. Remarks at Arms Control Association Annual Meeting, Washington, DC, May 10, 2011.

nine long days of floor deliberations engaged the Senate, especially its newer Members, in an extended seminar on the composition of our nuclear arsenal, the health of our stockpile, and the relationship between nuclear weapons and our national security. When the Senate voted for the Treaty, it inherently affirmed that our stockpile is safe, secure, and effective, and can be kept so without nuclear testing.

More importantly, the New START debate helped cultivate emerging new arms control champions, such as Senator [Jeanne] Shaheen [D-NH] and Senator [Bob] Casey [D-PA], who are here today. Before the debate, there was not a lot of muscle memory on treaties, especially nuclear treaties in the Senate. Now, there is. So we are in a stronger position to make the case for the CTBT on its merits. To maintain and enhance that momentum, the Obama Administration is preparing to engage the Senate and the public on an education campaign that we expect will lead to ratification of the CTBT.

In our engagement with the Senate, we want to leave aside the politics and explain why the CTBT will enhance our national security. Our case for Treaty ratification consists of three primary arguments.

One, the United States no longer needs to conduct nuclear explosive tests, plain and simple. Two, a CTBT that has entered into force will obligate other states not to test and provide a disincentive for states to conduct such tests. And three, we now have a greater ability to catch those who cheat.

Let me take these points one by one.

From 1945 to 1992, the United States conducted more than 1,000 nuclear explosive tests—more than all other nations combined. The cumulative data gathered from these tests have provided an impressive foundation of knowledge for us to base the continuing effectiveness of our arsenal. But historical test data alone is insufficient.

Well over a decade ago, we launched an extensive and rigorous Stockpile Stewardship program that has enabled our nuclear weapons laboratories to carry out the essential surveillance and warhead life extension programs to ensure the credibility of our deterrent.

Every year for the past 15 years, the Secretaries of Defense and Energy from Democratic and Republican Administrations, and the directors of the nuclear weapons laboratories have certified that our arsenal is safe, secure, and effective. And each year they have affirmed that we do not need to conduct explosive nuclear tests.

The lab directors tell us that Stockpile Stewardship has provided a deeper understanding of our arsenal than they ever had when testing was commonplace. Think about that for a moment. Our current efforts go a step beyond explosive testing by enabling the labs to anticipate problems in advance and reduce their potential impact on our arsenal—something that nuclear testing could not do. I, for one, would not trade our successful approach based on world-class science and technology for a return to explosive testing.

This Administration has demonstrated an unprecedented commitment to a safe, secure, and effective arsenal so long as nuclear weapons exist. Despite the narrative put forward by some, this Administration inherited an underfunded and underappreciated nuclear complex. We have worked tirelessly to fix that situation and ensure our complex has every asset needed to achieve its mission.

The President has committed $88 billion in funding over the next decade to maintain a modern nuclear arsenal, retain a modern nuclear weapons production complex, and nurture a highly trained workforce. At a time when every part of the budget is under the microscope, this pledge demonstrates our commitment and should not be discounted. To those who doubt our commitment, I ask them to put their doubts aside and invest the hard work to support our budget requests in the Congress.

When it comes to the CTBT, the United States is in a curious position. We abide by the core prohibition of the Treaty because we don't need to test nuclear weapons. And we have contributed to the development of the International Monitoring System. But the principal benefit of ratifying the Treaty, constraining other states from testing, still eludes us. That doesn't make any sense to me and it shouldn't make any sense to the Members of the Senate.

I do not believe that even the most vocal critics of the CTBT want to resume explosive nuclear testing. What they have chosen instead is a status quo where the United States refrains from testing without using that fact to lock in a legally binding global ban that would significantly benefit the United States.

Second, a CTBT that has entered into force will hinder other states from advancing their nuclear weapons capabilities. Were the CTBT to enter into force, states interested in pursuing or advancing a nuclear weapons program would risk either deploying weapons that might not work or incur international condemnation and sanctions for testing.

While states can build a crude first generation nuclear weapon without conducting nuclear explosive tests, they would have trouble going further, and they probably wouldn't even know for certain the yield of the weapon they built. More established nuclear weapons states could not, with any confidence, deploy advanced nuclear weapon capabilities that deviated significantly from previously tested designs without explosive testing.

Nowhere would these constraints be more relevant than in Asia, where you see states building up and modernizing their forces. A legally binding prohibition on all nuclear explosive testing would help reduce the chances of a potential regional arms race in the years and decades to come.

Finally, we have become very good at detecting potential cheaters. If you test, there is a very high risk of getting caught. Upon the Treaty's entry into force, the United States would use the International Monitoring System [IMS] to complement our own state of the art national technical means to verify the Treaty.

In 1999, not a single certified IMS station or facility existed. We understand why some senators had doubts about its future, untested capabilities. But today the IMS is more than 75 percent complete. Two hundred fifty-four of the planned 321 monitoring stations are in place and functioning. And 10 of 16 projected radio-nuclide laboratories have been completed. The IMS detected both of North Korea's two announced nuclear tests.

While the IMS did not detect trace radioactive isotopes confirming that the 2009 event was in fact a nuclear explosive test, there was sufficient evidence to support an on-site inspection. On-site inspections are only permissible once the Treaty enters into force. An on-site inspection could have clarified the ambiguity of the 2009 test.

While the IMS continues to prove its value, our national technical means remain second to none and we continue to improve them. Last week, our colleagues at the NNSA conducted the first of a series of Source Physics Experiments at the Nevada Nuclear Security Site. These experiments will allow the United States to validate and improve seismic models and the use of new generation technology to further monitor compliance with the CTBT. Senators can judge our overall capabilities for themselves by consulting the National Intelligence Estimate released last year.

Taken together, these verification tools would make it difficult for any state to conduct nuclear tests that escape detection. In other words, a robust verification regime carries an important deterrent value in and of itself. Could we imagine a far-fetched scenario where a country might conduct a test so low that it would not be detected? Perhaps. But could a country be certain that it would not be caught? That is unclear. Would a country be willing to risk being caught cheating? Doubtful, because there would be a significant cost to pay for those countries that test.

We have a strong case for Treaty ratification. In the coming months, we will build upon and flesh out these core arguments. We look forward to objective voices providing their opinions on this important issue. Soon, the National Academy of Sciences, a trusted and unbiased voice on scientific issues, will release an unclassified report examining the Treaty from a technical perspective. The report will look at how U.S. ratification would impact our ability to maintain our nuclear arsenal and our ability to detect and verify explosive nuclear tests.

Let me conclude by saying that successful U.S. ratification of the CTBT will help facilitate greater international cooperation on the other elements of the President's Prague Agenda. It will strengthen our leverage with the international community to pressure defiant regimes like those in Iran and North Korea as they engage in illicit nuclear activities. We will have greater credibility when encouraging other states to pursue nonproliferation objectives, including universality of the Additional Protocol.

In short, ratification helps us get more of what we want. We give up nothing by ratifying the CTBT. We recognize that a Senate debate over ratification will be spirited, vigorous, and likely contentious. The debate in 1999, unfortunately, was too short and too politicized. The Treaty was brought to the floor without the benefit of extensive Committee hearings or significant input from Administration officials and outside experts.

We will not repeat those mistakes.

But we will make a more forceful case when we are certain the facts have been carefully examined and reviewed in a thoughtful process. We are committed to taking a bipartisan and fact-based approach with the Senate.

For my Republican friends who voted against the Treaty and might feel bound by that vote, I have one message: Don't be. The times have changed. Stockpile Stewardship works. We have made significant advances in our ability to detect nuclear testing. As my good friend George Shultz [U.S. Secretary of State, 1982–1989 in the Reagan adminstration] likes to say, those who opposed the Treaty in 1999 can say they were right, but they would be right to vote for the Treaty today.

We have a lot of work to do to build the political will needed to ratify the CTBT. Nuclear testing is not a front-burner issue in the minds of most Americans, in part, because we have not tested in nearly 20 years. To understand the gap in public awareness, just think that in 1961 some 10,000 women walked off their job as mothers and housewives to protest the arms race and nuclear testing. Now, that strike did not have the same impact as the nonviolent marches and protests to further the cause of Civil Rights.

But the actions of mothers taking a symbolic and dramatic step to recognize global nuclear dangers showed that the issue has resonance beyond "the Beltway," beyond the think tank world and beyond the Ivory Tower. That level of concern is there today and we need your energy, your organizational skills, and your creativity to tap into it.

If we are to move safely and securely to a world without nuclear weapons, then we need to build the requisite political support and that can only be done by people like you.

Ellen Tauscher is the U.S. Under Secretary of State for arms control and international security. She previously served in the U.S. House of Representatives representing the Tenth Congressional District of California. She has been an investment banker and a member and officer of the New York Stock Exchange. Tauscher holds a BS degree from Seton Hall University.

Baker Spring

 **NO**

U.S. Should Reject Ratification of the Comprehensive Test Ban Treaty

The United States Senate voted to reject ratification of the 1996 Comprehensive Test Ban Treaty (CTBT) on October 13, 1999 [by a vote of 48 in favor to 51 opposed]. This determinate action by the Senate should have marked the end of consideration of the treaty by the U.S. Nevertheless, Under Secretary of State for Arms Control and International Security Ellen Tauscher recently told an audience that the Administration is preparing to engage the Senate and the public on an education campaign that is designed to lead to U.S. ratification of the CTBT.

The substantive problems that led to the Senate's considered judgment in 1999 remain relevant today. If anything, they have worsened in the intervening years. But procedurally, there is no justification for reconsideration of the treaty today. The institutional integrity of the Senate is now at stake.

Substantive Problems with the CTBT Persist

According to Tauscher, Senate consent to the ratification of the CTBT may be justified on the basis that "times have changed." In reality, the substantive problems with the CTBT that led to its rejection in 1999 are still present. In fact, the problems regarding the maintenance of a safe, reliable, and militarily effective nuclear arsenal have grown worse over the intervening years:

- **The CTBT does not define what it purports to ban.** The text of the treaty remains identical to that which the Senate rejected in 1999. Its central provision, as well as its object and purpose, is to ban explosive nuclear testing. The treaty does not, however, define the term. The U.S. interpretation is that it means a "zero-yield" ban, but other states may not share that interpretation.
- **The U.S. nuclear weapons complex has grown weaker during the intervening years.** After considerable pressure from a number of senators, chief among them Jon Kyl (R-AZ), about the alarming decline in the U.S.'s nuclear weapons, the [President Barack] Obama Administration committed to invest more money in the complex in order to pressure the Senate into granting consent to the badly flawed New START [Strategic Arms Reduction Treaty] arms control treaty with Russia. [The New START agreement was ratified by the U.S. Senate in December 2010.] But this investment program is only just getting started, and its success is far from guaranteed.

- **A zero-yield ban on nuclear explosive tests remains unverifiable.** If the U.S. interpretation of the CTBT as a zero-yield ban is accurate, it was impossible to verify the ban in 1999, and it remains so today. The International Monitoring System (IMS) being put in place to detect violations depends largely on seismic evidence. The fact is that extremely low-yield tests are not likely to be detected by the IMS. Even Tauscher acknowledged that it is possible that a "country might conduct a test so low [in yield] that it would not be detected." At the same time, she dismissed this possibility as "far-fetched." In reality, it is not at all far-fetched.

- **The Obama Administration has imposed self-defeating output limits on the nuclear weapons modernization program.** While the Obama Administration has pledged to increase the investment level in the nuclear weapons complex and stockpile stewardship programs, it is also imposing limits on what the complex and program may do. Specifically, the April 2010 Nuclear Posture Review Report [by the U.S. Department of Defense] states: "The United States will not develop new nuclear warheads. . . . Life Extension Programs will use only nuclear components based on previously tested designs, and will not support new military missions or provide for new military capabilities."

- **Nuclear proliferation trends are pointing in the wrong direction.** The Obama Administration sees its nuclear disarmament agenda, of which CTBT ratification is a part, as necessary to giving the U.S. the moral standing to combat nuclear proliferation. The fact that countries such as Iran, North Korea, and Pakistan are continuing to pursue or expand their nuclear weapons capabilities suggests that the Obama Administration's moral suasion argument is ineffective and that Iran and North Korea view the U.S. commitment to nuclear disarmament as a sign of weakness to be exploited.

Undermining the Senate's Institutional Integrity

Tauscher charged that the debate in the Senate in 1999 was too politicized and too short. Contrary to her assertion, the Senate's opponents of CTBT ratification did not fail to exercise due diligence in their review of CTBT at that time. They reviewed the treaty carefully and made considered arguments against ratification. Their arguments proved convincing to the Senate as a whole, and they prevailed overwhelmingly in the subsequent vote.

Regarding the time for consideration, it was CTBT proponents in the Senate that insisted on its immediate consideration in 1999. They effectively charged Senator Jesse Helms (R-NC), a leading opponent of ratification, with engaging in obstructionism over the matter. Further, the debate and vote on the CTBT in the Senate was conducted under a painstakingly worked out *unanimous* consent agreement. This is a far cry from the recent procedure for the consideration of New START, where proponents, having failed to achieve a unanimous consent agreement, simply rammed the treaty through by invoking cloture [thereby ending a filibuster by the treaty's opponents]. Clearly, the proponents of the CTBT now view the Senate's 1999 vote to reject CTBT as procedurally illegitimate only because they lost.

The Senate should not take such an attack on its integrity lightly. Members of the Senate, therefore, would be justified in sending a letter to President Obama making the following two requests:

1. **That President Obama ask the Senate to return the CTBT to the executive branch.** On the basis that the 1999 Senate vote to reject ratification of the CTBT was the Senate's considered and institutional judgment on the matter, President Obama should ask the Senate to terminate any further domestic consideration of the treaty's ratification by returning it to the executive branch. If President Obama fails to respond, the Senate may wish to consider returning the treaty to the executive branch on its own volition.

2. **That President Obama announce that the U.S. has no intention of ratifying the CTBT.** Senators may also ask President Obama to act in a manner consistent with Article 18 of the Vienna Convention on the Law of Treaties and announce that the U.S. has no intention of ratifying the CTBT. Not only would this relieve the U.S. of the obligation not to take actions contrary to the object and purpose of the treaty, but it would resolve the entire matter of CTBT entry into force. This is because Article XIV of the treaty requires that the U.S., among other states, become a party before it may enter into force. Senators, by making this request, would help bring the entire matter of the CTBT's entry into force to an appropriate conclusion.

Stand Up

The Senate's action to reject ratification in 1999 should be recognized and honored by its current members. Nothing has changed in the past 12 years to make the treaty any more palatable. Moreover, the integrity of the Senate as an institution is now being attacked. The Senate should not take this attack lying down.

Baker Spring is F. M. Kirby Research Fellow in National Security Policy in the Douglas and Sarah Allison Center for Foreign Policy Studies at the Heritage Foundation. He served as a defense and foreign policy expert on the staffs of U.S. Senators Paula Hawkins (R-FL) and David Karnes (R-NE). He received his master's degree in national security studies from Georgetown University.

EXPLORING THE ISSUE

Should the United States Ratify the Comprehensive Nuclear Test Ban Treaty?

Critical Thinking and Reflection

1. Neither side in the debate can be absolutely sure that what they advocate is the best path to peace and security. Is it better to err on the side of taking a further step to eliminating all nuclear arms or is it better to err on the side of "peace through strength," as President Ronald Reagan used to say?
2. There is an argument that nuclear weapons have been a force for peace because they have made war between the major powers too dangerous. Do you think that argument has any validity?
3. Do growing tensions between the Untitled States and China and the United States and Russia, each of which is an issue in this volume, make it more likely or less likely that you would support ratifying the CTBT or do these tension make no difference in your opinion?
4. Given the huge number of nuclear weapons that the United States has deployed or in reserve, does it make any difference in your view of the CTBT issue if countries like North Korea or Iran acquire a small nuclear arms inventory?

Is There Common Ground?

The possibility that the Senate might reverse its 1999 rejection of the CTBT has gone from dim to almost completely dark in response to the behavior of Russia in particular, and the growing strength of and tensions with China more generally. In late August 2014 during the crisis over Ukraine Putin rattled his nuclear arms, proclaiming, "I want to remind you that Russia is one of the most powerful nuclear nations. . . . This is a reality, not just words." Then just two weeks later, Russia tested a new version of its submarine-launched ballistic missiles, and, in a speech to military officers, Putin pointed to what he characterized as threats by the North Atlantic Treaty Organization and announced an increased effort to modernize Russia's strategic nuclear capabilities. "We have warned many times that we would have to take corresponding countermeasures to ensure our security," he told the audience. He added that the West should not get "hysterical" about Russia's intentions and claimed that a new nuclear arms race was "absolutely excluded." There have also been allegations that Russia has violated the Intermediate-Range Nuclear Forces Treaty (1987), by which President Ronald Reagan

and Soviet President Mikhail Gorbachev agreed to eliminate all their nuclear-capable missiles with ranges between 500 and 5,000 kilometers (about 300 to 3,000 miles).

The common reaction to these events is to oppose taking further steps to try to eliminate nuclear weapons even through such methods as barring non-nuclear armed countries from developing or testing them and for the countries that have nuclear weapons, restrictions on testing and decreasing limits on the number, types, and capabilities of their nuclear arsenals. Perhaps that is the wrong reaction, one that could create a self-fulfilling prophecy that will initiate a new and explosively dangerous arms race.

For now, a middle ground is the existing self-imposed moratorium on testing by all the major nuclear powers. The Senate may have rejected the CTBT in 1999, but during the 15 following years the United States has not conducted any tests. Nor have China or Russia. Only three nuclear explosive tests have occurred, all by North Korea, during these years.

Create Central

www.mhhe.com/createcentral

Additional Resources

Deibel, Terry L. (2002). "The Death of a Treaty." *Foreign Affairs* (81/5: 142–161)

Horovitz, Livlu. (2011). "A Detour Strategy for the Test Ban Treaty." *Washington Quarterly* (34/4: 87–99)

Inhofe, James. (September 8, 2014). "It's Time to Stop Putin's Nuclear Arms Buildup." *Foreign Policy* online at: http://www.foreignpolicy.com/articles /2014/09/08/its_time_to_stop_putins_nuclear_arms _buildup_inf_treaty

Medalla, Jonathan. (November 23, 2009). *Comprehensive Nuclear-Test-Ban Treaty: Background and Current Developments*. Congressional Research Service, Report RL33548

Internet References . . .

Comprehensive Nuclear Test-Ban Treaty (CTBT)—U.S. Department of State

http://www.state.gov/t/avc/c42328.htm

Fact Sheet: Global Nuclear Weapons Inventories in 2014

Published by the Center for Arms Control and Non-Proliferation

http://armscontrolcenter.org/issues/nuclearweapons /articles/fact_sheet_global_nuclear_weapons _inventories_in_2014/

Preparatory Commission for the Comprehensive Nuclear-Test-Ban Organization

http://www.ctbto.org/

United Nations Office of Disarmament Affairs

http://www.un.org/disarmament/WMD/Nuclear /CTBT.shtml

Selected, Edited, and with Issue Framing Material by:
John T. Rourke, *University of Connecticut, Storrs*

ISSUE

Does Using Drones to Attack Terrorists Globally Violate International Law?

YES: Mary Ellen O'Connell, from "Lawful Use of Combat Drones," Testimony During Hearings on "Rise of the Drones II: Examining the Legality of Unmanned Targeting," before the Subcommittee on National Security and Foreign Affairs, Committee on Oversight and Government Reform, U.S. House of Representatives (2010)

NO: Michael W. Lewis, from "Examining the Legality of Unmanned Targeting," Testimony During Hearings on "Rise of the Drones II: Examining the Legality of Unmanned Targeting," before the Subcommittee on National Security and Foreign Affairs, Committee on Oversight and Government Reform, U.S. House of Representatives (2010)

Learning Outcomes

After reading this issue, you will be able to:

- Understand what unmanned aerial vehicles (UVAs, drones) are and what their capabilities are.
- Discuss why UVAs are controversial weapons under international law.
- Comment on whether the upsurge of global terrorism has created a new realm of war that requires rethinking international way in many such areas as drones discussed here and the detention and treatment of terrorists, as discussed elsewhere in this volume.

ISSUE SUMMARY

YES: Mary Ellen O'Connell, a research professor at the Kroc Institute, University of Notre Dame, and the Robert and Marion Short Professor of Law at the School of Law, University of Notre Dame, tells a congressional committee that the United States is failing more often than not to follow the most important single rule governing drones: restricting their use to the battlefield.

NO: Michael W. Lewis, a professor of law at Ohio Northern University's Pettit College of Law, disagrees, contending that there is nothing inherently illegal about using drones to target specific terrorists or groups of terrorists on or away from the battlefield.

During March and April 2010, the Subcommittee on National Security and Foreign Affairs of the Committee on Oversight and Government Reform in the U.S. House of Representatives held a series of hearings to look into the military use of unmanned aerial vehicles (UAVs, drones). These remotely piloted aircraft are capable of launching missiles and otherwise attacking targets, of using cameras and other sensors to gather intelligence, of facilitating communications, and of performing other tasks. The characteristics of the numerous types of UAVs in the U.S. military inventory vary considerably by mission, but one of the best known is the Predator. It is propeller driven, has a top speed of 135 mph, has a 450-mile range, is 27 feet long, has a 48-foot wingspan, can stay aloft for 20 hours, and is armed with two air-to-surface Hellfire missiles, each tipped with a warhead carrying 20 pounds of high explosives. Relative to piloted warplanes, UAVs are inexpensive, costing about one-tenth as much each. There are also great differences in the time and money spent creating pilots.

During the first session of the hearings on March 23, 2010, the subcommittee's chair, Representative John F. Tierney (D-MA), opened the inquiry by outlining its purpose. As Tierney put it with regard to the subject of this debate:

> . . . Over the last decade, the number of unmanned systems and their applications has grown rapidly. So too has the number of operational, political, and legal questions associated with this technology. The growing demand for and reliance on unmanned systems has serious implications. . . . As the United States is engaged in two wars abroad, unmanned systems, particularly unmanned aerial vehicles, have become a centerpiece of that war effort. In recent years, the Department of Defense's UAV inventory has rapidly grown in size, from 167 in 2002 to over 7000 today. Last year, for the first time, the U.S. Air Force trained more unmanned pilots than traditional fighter pilots.
>
> Some express no doubt that unmanned systems have been a boost to U.S. war efforts in the Middle East and South Asia. CIA Director Leon Panetta said last May that "drone strikes are the only game in town in terms of confronting or trying to disrupt the al Qaeda leadership." Media reports over the last year that the top two leaders

of the Pakistani Taliban were killed by drone strikes also support this argument.

But some critics argue that drone strikes are unethical at best and counter-productive at worst. They point to the reportedly high rate of civilian casualties . . . and argue that the strikes do more to stoke anti-Americanism than they do to weaken our enemies. . . . This is particularly relevant in the era of counter-insurgency doctrine, a central tenet of which is, "first, do no harm."

It also may be the case that we are fighting wars with modern technology under an antiquated set of laws. For example, if the United States uses unmanned weapons systems, does that require an official declaration of war or an authorization for the use of force? . . .

These trends are already forcing us to ask new questions about domestic airspace regulation: who is allowed to own unmanned systems, and where they are allowed to operate them?

These are some of the questions that we will begin to answer in this hearing. Surely we will not conclude this conversation in one afternoon. . . .

In the following readings, two experts on international law relating to war take up the use of UAVs to attack targets away from an immediate war zone or "battlefield." Ellen O'Connell argues that such attacks violate international law. Michael Lewis disagrees.

YES ⤶

Mary Ellen O'Connell

Lawful Use of Combat Drones

Combat drones are battlefield weapons. They fire missiles or drop bombs capable of inflicting very serious damage. Drones are not lawful for use outside combat zones. Outside such zones, police are the proper law enforcement agents and police are generally required to warn before using lethal force. Restricting drones to the battlefield is the most important single rule governing their use. Yet, the United States is failing to follow it more often than not. At the very time we are trying to win hearts and minds to respect the rule of law, we are ourselves failing to respect a very basic rule: remote weapons systems belong on the battlefield.

I. A Lawful Battlefield Weapon

The United States first used weaponized drones during the combat in Afghanistan that began on October 7, 2001. We requested permission from Uzbekistan, which was then hosting the U.S. air base where drones were kept. We also used combat drones in the battles with Iraq's armed forces in the effort to topple Saddam Hussein's government that began in March 2003. We are still using drones lawfully in the ongoing combat in Afghanistan. Drones spare the lives of pilots, since the unmanned aerial vehicle is flown from a site far from the attack zone. If a drone is shot down, there is no loss of human life. Moreover, on the battlefield drones can be more protective of civilian lives than high aerial bombing or long-range artillery. Their cameras can pick up details about the presence of civilians. Drones can fly low and target more precisely using this information. [The U.S. commander in Afghanistan] General [Stanley] McChrystal has wisely insisted on zero-tolerance for civilian deaths in Afghanistan. The use of drones can help us achieve that. What drones cannot do is comply with police rules for the use of lethal force away from the battlefield. In law enforcement it must be possible to warn before using lethal force, in war-fighting this is not necessary, making the use of bombs and missiles lawful. The United Nations Basic Principles for the Use of Force and Firearms by Law Enforcement Officials

(*UN Basic Principles*) set out the international legal standard for the use of force by police:

> Law enforcement officials shall not use firearms against persons except in self-defense or defense of others against the imminent threat of death or serious injury, to prevent the perpetration of a particularly serious crime involving grave threat to life, to arrest a person presenting such a danger and resisting their authority, or to prevent his or her escape, and only when less extreme means are insufficient to achieve these objectives. In any event, intentional lethal use of firearms may only be made when strictly unavoidable in order to protect life.

The United States has failed to follow these rules by using combat drones in places where no actual armed conflict was occurring or where the U.S. was not involved in the armed conflict. On November 3, 2002, the CIA used a drone to fire laser-guided Hellfire missiles at a passenger vehicle traveling in a thinly populated region of Yemen. At that time, the Air Force controlled the entire drone fleet, but the Air Force rightly raised concerns about the legality of attacking in a place where there was no armed conflict. CIA agents based in Djibouti carried out the killing. All six passengers in the vehicle were killed, including an American. In January 2003, the United Nations Commission on Human Rights received a report on the Yemen strike from its special rapporteur on extrajudicial, summary, or arbitrary killing. The rapporteur concluded that the strike constituted "a clear case of extrajudicial killing."

Apparently, Yemen gave tacit consent for the strike. States [countries] cannot, however, give consent to a right they do not have. States may not use military force against individuals on their territory when law enforcement measures are appropriate. At the time of the strike, Yemen was not using military force anywhere on its territory. More recently, Yemen has been using military force to suppress militants in two parts of the country. The U.S.'s ongoing drone use, however, has not been part of those campaigns.

The United States has also used combat drones in Somalia probably starting in late 2006 during the Ethiopian invasion when the U.S. assisted Ethiopia in its attempt to install a new government in that volatile country. Ethiopia's effort had some support from the UN and the African Union. To the extent that the U.S. was assisting Ethiopia, our actions had some justification. It is clear, however, that the U.S. has used drone strikes independently of the attempt to restore order in Somalia. The U.S. has continued to target and kill individuals in Somalia following Ethiopia's pullout from the country.

The U.S. use of drones in Pakistan has similar problems to the uses in Yemen and Somalia. Where military force *is* warranted to address internal violence, governments have widely resorted to the practice of inviting in another state to assist. This is the legal justification the U.S. cites for its use of military force today in Afghanistan and Iraq. Yet, the U.S. cannot point to invitations from Pakistan for most of its drone attacks. Indeed, for much of the period that the United States has used drones on the territory of Pakistan, there has been no armed conflict. Therefore, even express consent by Pakistan would not justify their use.

The United States has been carrying out drone attacks in Pakistan since 2004. Pakistani authorities only began to use major military force to suppress militancy in May 2009, in Buner Province. Some U.S. drone strikes have been coordinated with Islamabad's efforts, but some have not. Some strikes have apparently even targeted groups allied with Islamabad.

II. The Battlefield Defined

The Bush administration justified the 2002 Yemen strike and others as justified under the law of armed conflict in the "Global War on Terror." The current State Department Legal Adviser, Harold Koh, has rejected the term "Global War on Terror," preferring to base our actions on the view that the U.S. is in an "armed conflict with al-Qaeda, the Taliban and associated forces." Under the new label, the U.S. is carrying out many of the same actions as the Bush administration under the old one: using lethal force without warning, far from any actual battlefield.

Armed conflict, however, is a real thing. The United States is currently engaged in an armed conflict in Afghanistan. The United States has tens of thousands of highly trained troops fighting a well-organized opponent that is able to hold territory. The situation in Afghanistan today conforms to the definition of armed conflict in international law. The International Law Association's Committee on the Use of Force issued a report in 2008 confirming the basic characteristics of all armed conflict: (1) the presence of organized armed groups that

are (2) engaged in intense inter-group fighting. The fighting or hostilities of an armed conflict occur within limited zones, referred to as combat or conflict zones. It is only in such zones that killing enemy combatants or those taking a direct part in hostilities is permissible.

Because armed conflict requires a certain intensity of fighting, the isolated terrorist attack, regardless of how serious the consequences, is not an armed conflict. Terrorism is crime. Members of al Qaeda or other terrorist groups are active in Canada, France, Germany, Indonesia, Morocco, Saudi Arabia, Spain, the United Kingdom, Yemen and elsewhere. Still, these countries do not consider themselves in a war with al Qaeda. In the words of a leading expert on the law of armed conflict, the British Judge on the International Court of Justice, Sir Christopher Greenwood:

> In the language of international law there is no basis for speaking of a war on Al-Qaeda or any other terrorist group, for such a group cannot be a belligerent, it is merely a band of criminals, and to treat it as anything else risks distorting the law while giving that group a status which to some implies a degree of legitimacy.

To label terrorists "enemy combatants" lifts them out of the status of *criminal* to that of *combatant,* the same category as America's own troops on the battlefield. This move to label terrorists combatants is contrary to strong historic trends. From earliest times, governments have struggled to prevent their enemies from approaching a status of equality. Even governments on the verge of collapse due to the pressure of a rebel advance have vehemently denied that the violence inflicted by their enemies was anything but criminal violence. Governments fear the psychological and legal advantages to opponents of calling them "combatants" and their struggle a "war."

President Ronald Reagan strongly opposed labeling terrorists combatants. He said that to "grant combatant status to irregular forces even if they do not satisfy the traditional requirements . . . would endanger civilians among whom terrorists and other irregulars attempt to conceal themselves."

The United Kingdom and other allies take the same position as President Reagan: "It is the understanding of the United Kingdom that the term 'armed conflict' of itself and in its context denotes a situation of a kind which is not constituted by the commission of ordinary crimes including acts of terrorism whether concerted or in isolation."

In the United States and other countries plagued by al Qaeda, institutions are functioning normally. No one has declared martial law. The International Committee of the Red Cross is not active. Criminal trials of suspected terrorists are being held in regular criminal courts. The police use lethal force only in situations of necessity. The U.S.'s actions today are generally consistent with its long-term policy of separating acts of terrorism from armed conflict—except when it comes to drones.

III. Battlefield Restraints

Even when the U.S. is using drones at the request of Pakistan in battles it is waging, we are failing to follow important battlefield rules. The U.S. must respect the principles of necessity, proportionality and humanity in carrying out drone attacks. "Necessity" refers to military necessity, and the obligation that force is used only if necessary to accomplish a reasonable military objective. "Proportionality" prohibits that "which may be expected to cause incidental loss of civilian life, injury to civilians, damage to civilian objects, or a combination thereof, which would be excessive in relation to concrete and direct military advantage anticipated." These limitations on permissible force extend to both the quantity of force used and the geographic scope of its use.

Far from suppressing militancy in Pakistan, drone attacks are fueling the interest in fighting against the United States. This impact makes the use of drones difficult to justify under the terms of military necessity. Most serious of all, perhaps, is the disproportionate impact of drone attacks. A principle that provides context for all decisions in armed conflict is the principle of humanity. The principle of humanity supports decisions in favor of sparing life and avoiding destruction in close cases under either the principles of necessity or proportionality. According to the International Committee of the Red Cross, the principles of necessity and humanity are particularly important in situations such as Pakistan:

> In classic large-scale confrontations between well-equipped and organized armed forces or groups, the principles of military necessity and of humanity are unlikely to restrict the use of force against legitimate military targets beyond what is already required by specific provisions of IHL [international humanitarian law]. The practical importance of their restraining function will increase with the ability of the conflict to control the circumstances and area in which its military operations are conducted, may become decisive where armed forces operate against selected individuals in situations comparable to peacetime policing. In practice, such considerations are likely to become particularly relevant where a party to the conflict exercises effective territorial control, most notably in occupied territories and non-international armed conflicts.

Another issue in drone use is the fact that strikes are carried out in Pakistan by the CIA and civilian contractors. Only members of the United States armed forces have the combatant's privilege to use lethal force without facing prosecution. CIA operatives are not trained in the law of armed conflict. They are not bound by the Uniform Code of Military Justice to respect the laws and customs of war. They are not subject to the military chain of command. This fact became abundantly clear during the revelation of U.S. use of harsh interrogation tactics. Given the negative impact of that unlawful conduct on America's standing in the world and our ability to promote the rule of law, it is difficult to fathom why the Obama administration is using the CIA to carry out drone attacks, let alone civilian contractors.

Conclusion

The use of military force in counter-terrorism operations has been counterproductive. Military force is a blunt instrument. Inevitably unintended victims are the result of almost any military action. Drone attacks in Pakistan have resulted in large numbers of deaths and are generally seen as fueling terrorism, not abating it. In congressional testimony in March 2009, counter-terrorism expert, David Kilcullen, said drones in Pakistan are giving "rise to a feeling of anger that coalesces the population around the extremists and leads to spikes of extremism well outside the parts of the country where we are mounting those attacks." Another expert told the *New York Times*, "The more the drone campaign works, the more it fails—as increased attacks only make the Pakistanis angrier at the collateral damage and sustained violation of their sovereignty." A National Public Radio Report on April 26, 2010, pointed out that al Qaeda is losing support in the Muslim world because of its violent, lawless tactics. We can help eliminate the last of that support by distinguishing ourselves through commitment to the rule of law, especially by strict compliance with the rules governing lethal force.

MARY ELLEN O'CONNELL is the research professor of international dispute resolution at the Kroc Institute, the University of Notre Dame, and also the Robert and Marion Short professor of law at the university's law school. Among her publications is *The Power and Purpose of International Law* (Ohio State University Press, 2008). She earned her JD from Columbia University.

Michael W. Lewis  **NO**

Examining the Legality of Unmanned Targeting

Introduction

I am a professor of law at Ohio Northern University's Pettit College of Law where I teach International Law and the Law of Armed Conflict. I spent over 7 years in the U.S. Navy as a Naval Flight Officer flying F-14s. I flew missions over the Persian Gulf and Iraq as part of Operations Desert Shield/Desert Storm and I graduated from Top Gun [the U.S. Navy Fighter Weapons School] in 1992. After my military service I attended Harvard Law School and graduated *cum laude* in 1998. Subsequently I have lectured on a variety of aspects of the laws of war, with an emphasis on aerial bombardment, at dozens of institutions including Harvard, NYU, Columbia and the University of Chicago. I have published several articles and co-authored a book on the laws of war relating to the war on terror. My prior experience as a combat pilot and strike planner provides me with a different perspective from most other legal scholars on the interaction between law and combat.

The Current Laws of War Are Sufficient to Address the Drone Question

There is nothing inherently illegal about using drones to target specific individuals. Nor is there anything legally unique about the use of unmanned drones as a weapons delivery platform that requires the creation of new or different laws to govern their use.

As with any other attack launched against enemy forces during an armed conflict, the use of drones is governed by International Humanitarian Law (IHL). Compliance with current IHL that governs aerial bombardment and requires that all attacks demonstrate military necessity and comply with the principle of proportionality is sufficient to ensure the legality of drone strikes. In circumstances where a strike by a helicopter or an F-16

[a U.S. warplane] would be legal, the use of a drone would be equally legitimate. However, this legal parity does not answer three fundamental questions that have been raised by these hearings. Who may be targeted? Where may they be targeted? And finally who is allowed to pilot the drones and determine which targets are legally appropriate?

Who May Be Targeted?

In order to understand the rules governing the targeting of individuals, it is necessary to understand the various categories that IHL assigns to individuals. To best understand how they relate to one another it is useful to start from the beginning.

All people are civilians and are not subjected to targeting unless they take affirmative steps to either become combatants or to otherwise lose their civilian immunity. It is important to recognize that a civilian does not become a combatant by merely picking up a weapon. In order to become a combatant an individual must be a member of the "armed forces of a Party to a conflict." This definition is found in Article 43 of Additional Protocol I to the Geneva Conventions. It goes on to define the term "armed forces" as:

> The armed forces of a Party to a conflict consist of all organized armed forces, groups and units which are under a command responsible to that Party for the conduct of its subordinates, even if that Party is represented by a government or an authority not recognized by an adverse Party. Such armed forces shall be subject to an internal disciplinary system which, *inter alia*, shall enforce compliance with the rules of international law applicable in armed conflict.

The status of combatant is important because combatants "have the right to participate directly in hostilities." This "combatants' privilege" allows privileged individuals to participate in an armed conflict without violating domestic laws prohibiting the destruction of property, assault, murder, etc. The combatant's conduct is therefore regulated by IHL rather than [by] domestic law.

Combatant status is something of a double-edged sword, however. While it bestows the combatant privilege on the individual, it also subjects that individual to attack at any time by other parties to the conflict. A combatant may be lawfully targeted whether or not they pose a current threat to their opponents, whether or not they are armed, or even awake. The only occasion on which IHL prohibits attacking a combatant is when that combatant has surrendered or been rendered *hors de combat*. Professor Geoff Corn has argued compellingly that this ability to target based upon status, rather than on the threat posed by an individual, is the defining feature of an armed conflict.

After examining the definition of combatant, it becomes apparent that combatant status is based upon group conduct, not individual conduct. Members of al Qaeda are not combatants because as a group they are not "subject to an internal disciplinary system which [enforces] compliance with the rules of international law applicable in armed conflict." It does not matter whether an individual al Qaeda member may have behaved properly; he can never obtain the combatants' privilege because the group he belongs to does not meet IHL's requirements. Professor [David] Glazier's testimony [before this committee] that al Qaeda and the Taliban could possess "the basic right to engage in combat against us" is mistaken. These groups have clearly and unequivocally forfeited any "right" to be treated as combatants by choosing to employ means and methods of warfare that violate the laws of armed conflict, such as deliberately targeting civilians.

If al Qaeda members are not combatants, then what are they? They must be civilians, and civilians as a general rule are immune from targeting. However, civilians lose this immunity "for such time as they take a direct part in hostilities." The question of what constitutes direct participation in hostilities (DPH) has been much debated. While DOD [the U.S. Department of Defense] has yet to offer its definition of DPH, the International Committee of the Red Cross (ICRC) recently completed a six-year study on the matter and has offered interpretive guidance that, while not binding on the United States, provides a useful starting point. The ICRC guidance states that "members of organized armed groups [which do not qualify as combatants] belonging to a party to the conflict lose protection against direct attack for the duration of their membership (i.e., for as long as they assume a continuous combat function)."

The concept of a "continuous combat function" within DPH is a reaction to the "farmer by day, fighter by night" tactic that a number of organized armed terrorist groups have employed to retain their civilian immunity from attack for as long as possible. Because such individuals (be they fighters, bomb makers, planners or leaders) perform a continuous combat function, they may be directly targeted for as long as they remain members of the group. The only way for such individuals to reacquire their civilian immunity is to disavow membership in the group.

So the answer to "Who may be targeted?" is any member of al Qaeda or the Taliban, or any other individuals that have directly participated in hostilities against the United States. This would certainly include individuals that directly or indirectly (e.g. by planting IEDs) [improvised explosive devices] attacked Coalition forces as well as any leadership within these organizations. Significantly, the targeting of these individuals does not involve their elevation to combatant status as Professor O'Connell implied in her testimony [found in the first reading]. These individuals are civilians who have forfeited their civilian immunity by directly participating in hostilities. They are not, and cannot become, combatants until they join an organized armed group that complies with the laws of armed conflict, but they nevertheless remain legitimate targets until they clearly disassociate themselves from al Qaeda or the Taliban.

Where May Attacks Take Place?

Some witnesses have testified to this subcommittee that the law of armed conflict only applies to our ongoing conflict with al Qaeda in certain defined geographic areas. Professor O'Connell states that the geographic limit of the armed conflict is within the borders of Afghanistan while others include the border areas of Pakistan and Iraq. They take the position that any operations against al Qaeda outside of this defined geography are solely the province of law enforcement, which requires that the target be warned before lethal force is employed. Because drones cannot meet this requirement they conclude that drone strikes outside of this geographical area should be prohibited. The geographical boundaries proposed are based upon the infrequency of armed assaults that take place outside of Afghanistan, Iraq and the border region of Pakistan. Because IHL does not specifically address the geographic scope of armed conflicts, to assess these proposed requirements it is necessary to step back and consider the law of armed conflict as a whole and the realities of warfare as they apply to this conflict.

One of the principal goals of IHL is to protect the civilian population from harm during an armed conflict.

To further this goal IHL prohibits direct attacks on civilians and requires that parties to the conflict distinguish themselves from the civilian population. As a result, it would seem anomalous for IHL to be read in such a way as to reward a party that regularly targets civilians, and yet that is what is being proposed. As discussed above, a civilian member of al Qaeda who is performing a continuous combat function may be legitimately targeted with lethal force without any warning. But the proposed geographic limitations on IHL's application offer this individual a renewed immunity from attack. Rather than disavowing an organization that targets civilians, IHL's preferred result, the proposed geographic restrictions allow the individual to obtain the same immunity by crossing an international border and avoiding law enforcement while remaining active in an organization that targets civilians. When law enforcement's logistical limitations are considered, along with the host state's ambivalence for actively pursuing al Qaeda within its borders, it becomes clear that the proposed geographical limitations on IHL are tantamount to the creation of a safe haven for al Qaeda.

More importantly these proposed limitations would hand the initiative in this conflict over to al Qaeda. Militarily the ability to establish and maintain the initiative during a conflict is one of the most important strategic and operational advantages that a party can possess. To the extent that one side's forces are able to decide when, where and how a conflict is conducted, the likelihood of a favorable outcome is greatly increased. If IHL is interpreted to allow al Qaeda's leadership to marshal its forces in Yemen or the Sudan, or any number of other places that are effectively beyond the reach of law enforcement and to then strike at its next target of choice, whether it be New York, Madrid, London, Bali, Washington, DC or Detroit, then IHL is being read to hand the initiative in the conflict to al Qaeda. IHL should not be read to reward a party that consistently violates IHL's core principles and as Professor Glazier points out in his reference to the Cambodian incursion, it was not read that way in the past.

Those opposed to the position that IHL governs the conflict with al Qaeda regardless of geography, and therefore allows strikes like the one conducted in Yemen in 2002, have voiced three main concerns. The first concern is that the United States may be violating the sovereignty of other nations by conducting drone strikes on their territory. It is true that such attacks may only be conducted with the permission of the state on whose territory the attack takes place and questions have been raised about whether Pakistan, Yemen and other states have consented to this use of force. This is a legitimate concern that must be satisfactorily answered while accounting for the obvious sensitivity associated with granting such permission. The fact that Harold Koh, the State Department's Legal Advisor, specifically mentioned [in his testimony] the "sovereignty of the other states involved" in his discussion of drone strikes is evidence that the Administration takes this requirement seriously.

The second concern is that such a geographically unbounded conflict could lead to drone strikes in Paris or London, or to setting the precedent for other nations to employ lethal force in the United States against its enemies that have taken refuge here. These concerns are overstated. The existence of the permission requirement mentioned above means that any strikes conducted in London or Paris could only take place with the approval of the British or French governments. Further, any such strike would have to meet the requirements of military necessity and proportionality and it is difficult to imagine how these requirements could be satisfactorily met in such a congested urban setting.

Lastly, there is a legitimate concern that mistakes could be made. An individual could be inappropriately placed on the list and killed without being given any opportunity to challenge his placement on the list. Again, Mr. Koh's assurances that the procedures for identifying lawful targets "are extremely robust" are in some measure reassuring, particularly given his stature in the international legal community. However, some oversight of these procedures is clearly warranted. While *ex ante* [before the fact/event] review must obviously be balanced against secrecy and national security concerns, *ex post* review can be more thorough. When the Israeli Supreme Court approved the use of targeted killings, one of its requirements was for transparency after the fact coupled with an independent investigation of the precision of the identification and the circumstances of the attack. A similar *ex post* transparency would be appropriate here to ensure that "extremely robust" means something.

Who May Do the Targeting?

Another question raised in the hearings was the propriety of allowing the CIA to control drone strikes. Professor Glazier opined that CIA drone pilots conducting strikes are civilians directly participating in hostilities and suggested that they might be committing war crimes by engaging in such conduct. Even if these are not considered war crimes, if the CIA members are civilians performing a continuous combat function then they are not entitled to the combatants' privilege and could potentially be liable for domestic law violations.

Therefore, if CIA members are going to continue piloting drones and planning strikes, then they must obtain combatant status. Article 43(3) of Protocol I [to the Geneva Convention of 1949] allows a party to "incorporate a paramilitary or armed law enforcement agency into its armed forces" after notifying other parties to the conflict. For such an incorporation to be effective a clear chain of command would have to be established (if it does not already exist) that enforces compliance with the laws of armed conflict. Without this incorporation or some other measure clearly establishing the CIA's accountability for law of armed conflict violations, the continued use of CIA drone pilots and strike planners will be legally problematic.

Conclusion

Drones are legitimate weapons platforms whose use is effectively governed by current IHL applicable to aerial bombardment. Like other forms of aircraft they may be used to target enemy forces, whether specifically identifiable individuals or armed formations.

IHL permits the targeting of both combatants and civilians that are directly participating in hostilities. Because of the means and methods of warfare that they employ, al Qaeda and Taliban forces are not combatants and are not entitled to the combatants' privilege. They are instead civilians that have forfeited their immunity because of their participation in hostilities. Members of al Qaeda and the Taliban that perform continuous combat functions may be targeted at any time, subject to the standard requirements of distinction and proportionality.

Placing blanket geographical restrictions on the use of drone strikes turns IHL on its head by allowing individuals an alternative means for reacquiring effective immunity from attack without disavowing al Qaeda and its methods of warfare. It further bolsters al Qaeda by providing them with a safe haven that allows them to regain the initiative in their conflict with the United States. The geographical limitations on drone strikes imposed by sovereignty requirements, along with the ubiquitous requirements of distinction and proportionality, are sufficient to prevent these strikes from violating international law. However, some form of *ex post* [after the fact/event] transparency and oversight is necessary to review the identification criteria and strike circumstances to ensure that they remain "extremely robust."

Michael W. Lewis is a professor of law at Ohio Northern University's Pettit College of Law, where he teaches and writes in the fields of international law and the law of armed conflict. He has served as a U.S. Navy fighter pilot and holds a JD from Harvard University.

EXPLORING THE ISSUE

Does Using Drones to Attack Terrorists Globally Violate International Law?

Critical Thinking and Reflection

1. If al Qaeda were able to attack the White House using a small airplane with suicide bombers that killed Barack, Michelle, Sasha, and Malia Obama, would that strike be different in terms of *jus in bello* than the missile strike that specifically targeted Mustafa Abu al-Yazid and also killed his family members?
2. What do you think of the implication in the remarks of UN investigator Philip Alston that the ability to kill from a far distance, including now using robotic weapons system, is causing a "disconnect" from the death and suffering caused?
3. In April 2010, President Obama reportedly authorized a drone attack, if possible, on an American citizen living in Yemen. The individual, Muslim cleric Anwar al-Awlaki, is alleged, but not proven, to have been involved in several terrorist plots against the United States. Do you support President Obama's decision, and, if so, are there other circumstances in which the president can justifiably authorize the *de facto* execution of an American without judicial sanction?

Is There Common Ground?

Drone attacks have in a few short years gone from unusual to commonplace. The large majority of these attacks by U.S. drones have been against Taliban and al Qaeda targets in Pakistan. There was one attack in 2004 and three in 2005. In 2010, a total of 122 drone attacks were launched, and by the end of 2013, 381 attacks had taken place. There is no doubt that attacks in Pakistan and elsewhere have killed many fearsome terrorists. A September 2014 attack in Somalia, for instance, killed Ahmed Abdi Godane, the leader of the terrorist group Al Shabab, and several other group members. One indication of his importance was that the United States had offered a $7 million reward to his capture. But the death of such "high-value" targets and many more lower-value but lethal terrorists and supporters has been accompanied by many noncombatant deaths. The London-based Bureau of Investigative Journalism estimates that these attacks claimed somewhere between 2,537 and 3,646 lives. Of these, at least 25 percent and probably many more, of the dead have been innocent bystanders. Other studies have found both significantly higher and lower percentages of noncombatant casualties.

Inasmuch as zero-collateral damage in all military operations is neither an achievable standard nor required under *jus in bello* (just conduct of war) by international law, the issue of collateral damage remains highly debatable. This is especially true because the estimates of civilians killed and wounded vary widely. So do the descriptions of the boundaries of the battlefield. President Obama has authorized several increases in the geographic use of drones and permissible targets, and the State Department's legal adviser has defended those extensions, arguing that the war on terrorism is global, not confined to Afghanistan or any other battle zone and that the United States has a right of self-defense to attack terrorists and their supporters anywhere. The adviser also described U.S. procedures for identifying and hitting legal targets as "extremely robust" and becoming "even more precise." A different view was given to the United Nations Human Rights Council in a June 2010 report by UN special investigator. He criticized U.S. UAV attacks as having displaced "clear legal standards with a vaguely defined license to kill," and also for creating a "risk of developing a 'Playstation' mentality to killing."

Yet another dimension to the controversy broke out when a U.S. drone strike in Yemen targeted and killed an American citizen, Anwar al-Awlaki. Born in New Mexico, al-Awlaki had become a militant, gone to Yemen, and become an al Qaeda leader there. Civil rights activists protested that Obama had violated the Constitution by putting any American citizen on a list that the U.S. military and intelligence forces were allowed to seek out and kill.

The common ground for the use of drones is probably very similar to the standards of any form of warfare. Strikes should be launched only against legitimate enemy targets and care should be taken to avoid or minimize civilian casualties and other collateral damage. The practice, however, is much harder than the principle.

Create Central

www.mhhe.com/createcentral

Additional Resources

Brooks, Rosa. (2014). "Drones and the International Rule of Law." *Ethics & International Affairs* (28/1: 83–103)

Byman, Daniel. (2013). "Why Drones Work: The Case for Washington's Weapon of Choice." *Foreign Affairs* (92: 32–43)

Salinas de Frias, Ana Maria, Samuel, Katja, and White, Nigel, eds. (2012). *Counter-Terrorism: International Law and Practice.* (Oxford University Press)

Law of War Handbook. (2005). (U.S. military's Judge Advocate General's Legal Center and School). Accessed at: http://www.loc.gov/rr/frd/Military_Law/pdf/law-war-handbook-2005.pdf

Internet References . . .

The Laws of War

Part of the Avalon Project of Yale University.

http://avalon.law.yale.edu/subject_menus/lawwar.asp

War and Law

https://www.icrc.org/en/war-and-law

Selected, Edited, and with Issue Framing Material by:
John T. Rourke, *University of Connecticut, Storrs*

ISSUE

Is Closing the Prison for Alleged Terrorists at the U.S. Naval Base, Guantanamo, Cuba Justified?

YES: Elisa Massimino, from "Testimony During Hearings on 'Closing Guantanamo: The National Security, Fiscal, and Human Rights Implications,'" The Subcommittee on the Constitution, Civil Rights and Human Rights, Committee on the Judiciary, U.S. Senate (2013)

NO: Frank J. Gaffney, Jr., from "Testimony During Hearings on 'Closing Guantanamo: The National Security, Fiscal, and Human Rights Implications,'" The Subcommittee on the Constitution, Civil Rights and Human Rights, Committee on the Judiciary, U.S. Senate (2013)

Learning Outcomes

After reading this issue, you will be able to:

- Understand the difference that the U.S. government makes between "enemy combatants" and "prisoners of war" and how each group is dealt with by the United States.
- Evaluate the charge that many or even all the prisoners at Guantanamo Bay are being denied such basic rights as the right to be charged, to have a trial, and to see the evidence against them.
- Evaluate the debate that most prisoners at Guantanamo Bay are not dangerous.
- Explain the domestic legal and political problems President Obama faces to fulfilling his pledge to close the detention center at Guantanamo Bay.

ISSUE SUMMARY

YES: Elisa Massimino, president of Human Rights First, a U.S.-based advocacy organizations urging U.S. global leadership on human rights, tells Congress that there are tough questions to resolve regarding closing the U.S. prison at the Guantanamo Bay Naval Base, but that it is possible to close the facility with smart and sustained leadership from the president and Congress.

NO: Frank J. Gaffney, Jr., president of the Center for Security Policy, a private analysis and advocacy organization, maintains that the United States is at war with terrorism and in that struggle Guantanamo Bay is the optimal location for U.S. detention and interrogation of terrorists and associated unlawful enemy combatants.

The U.S. Naval Base at Guantanamo Bay, which houses the detention center at issue in this debate, is something of an anomaly. Located on the southern coast of Cuba near the eastern end of the island, Gitmo, as the 45-square mile base is often called, is a relic of both colonialism and the Cold War. Whatever the legalities of Cuban sovereignty, the United States first acquired possession of the bay and its surrounding area during the Spanish-American War of 1898. One result of Spain's defeat was that Cuba achieved independence in 1902. It was only a limited independence, though, because as a condition for the withdrawal of their troops, the Americans insisted that Cuba's new constitution grant the United States the right to supervise its finances and foreign affairs. Washington also required that Cuba lease land for a Guantanamo Bay to the United States. That was extended in 1934 by a U.S–Cuba treaty to include a perpetual lease.

After the revolution that brought Fidel Castro's communist government to power in 1959, he argued that the leases in 1903 and the 1934 treaty were void because they had been imposed on Cuba in a patently unequal relationship with the powerful United States. Castro certainly had at least an arguable point, but the politics of the Cold War and other factors have made it certain that successive U.S. president from Dwight Eisenhower to Barack Obama have maintained the U.S. claim to Gitmo. The yearly lease is $4,085. Cuba does not cash the checks. The base has never played a substantial role in U.S. military operations, and in an era of aircraft carriers and other long range weaponry, its role in U.S. national security is marginal at best. The less than 10,000 U.S. military personnel stationed there are needed more to operate the facilities and guard against a Cuban takeover.

The status of Gitmo rose from a national security backwater to a matter of national debate after 2001 as a result of the 9/11 attacks on the United States and the ensuing war on terrorism, including the U.S. invasion of Afghanistan. During the operations there and elsewhere the United States and its allies captured a substantial number of al Qaeda members and other alleged terrorists. This created a problem. The treatment of another country's military prisoners of war (POWs), including captured members of irregular forces such as guerillas, is governed by a series of treaties collectively called the Geneva Conventions (1864, 1929, 1949, and 1949). Generally, POWs may not be tried or punished for the acts they have committed in combat. The treatment of "home-grown" terrorism and terrorists (those who commit terrorist acts in their own country and have no global ties) is subject to their country's laws, and such terrorists are tried in their country's general court system. Captured international terrorists, arguably even including those who commit their acts of terrorist within their own country at the behest of or through association with an international terrorist organization, occupy an uncertain status between the clearly defined spheres of prisoners of war and domestic criminals.

The U.S. decision was to send most captured al Qaeda members and other alleged terrorists to the U.S. naval base at Guantanamo Bay, Cuba to be housed at a detention facility opened in January 2002. The administration of President George W. Bush argued that these prisoners were not members of an organized military and set up military commissions (tribunals) to try them. Defendants at Gitmo being tried by military commissions did not have many of the protections available to anyone, not just Americans citizens, held in U.S. prisons or tried in U.S. civilian courts or POWs tried in military courts martial. This difference

was challenged by one prisoner, and the Supreme Court ruled in *Hamden v. Rumsfeld* (2006) that the president had exceeded his authority by unilaterally establishing the military commissions. But Congress closed that defense by passing the Military Commission Act (MCA) in 2006 allowing the special tribunals.

As a candidate for president in 2008, Barack Obama deplored the use of military commissions as a "dangerously flawed legal approach," and he also promised to end the confinement of terrorists at Guantanamo Bay and transfer them to prisons in the United States. On his third day as president, at first, Obama moved to fulfill his campaign pledges but strong public opposition, a law passed by Congress barring bringing prisoners from Gitmo into the United States, and other difficulties soon led Obama to renew the use of military commissions and continue the operations of the prison.

Since 2002, a total of 779 prisoners have been held at Gitmo, some since the prison opened. Most of the 779 have been released or transferred to other countries for detention. This includes the five prisoners swapped in 2014 for the release of Bowe Bergdahl, a U.S. soldier captured in Afghanistan by the Taliban. Nine prisoners have died, most by suicide. Only eight have been tried and convicted by a military commission and another by a regular federal court. As of late 2012, 149 prisoners remain in custody. Of these, about half will be released if another country will accept them. The rest are presumably awaiting trial or are considered so dangerous that the United States refuses to release them even though (to protect intelligence sources) it lacks evidence that it is willing to present in court to convict them. There have been widespread accusations that some detainees have been subject to waterboarding and other treatment that critics charge range from abuse to torture. There is little doubt that at least some of the charges are accurate, although the U.S. government maintains that many of the charges are untrue, overwrought, and/or do not constitute torture.

In the YES selection that follows, Elisa Massimino contends that Guantanamo is a symbol for many around the world of torture, injustice, and illegitimacy, that Americans as a nation cannot be separated from what they do, that the prison should be closed, and that the detainees should be tried and convicted or released. Frank J. Gaffney, Jr., argues that not only is it desirable and necessary to continue to incarcerate detainees at Guantanamo Bay, but the United States should add to their number if that will help gather vital intelligence and keep dangerous jihadist enemy combatants off the battlefield.

YES

Elisa Massimino

Testimony During Hearings on "Closing Guantanamo: The National Security, Fiscal, and Human Rights Implications"

Introduction

My name is Elisa Massimino. I am the President and Chief Executive Officer of Human Rights First. Human Rights First is one of the nation's leading advocacy and action organizations. We believe that upholding human rights is not only a moral obligation; it is a vital national interest. America is strongest when our policies and actions match our values. For 35 years, Human Rights First has worked to ensure that our country is a beacon on human rights in a world that sorely needs American leadership.

Since 9/11, much of our work has focused on ensuring that our country stays true to its values, even as it confronts the threat of terrorism. We know from our 35 years of work around the world that what the United States does—particularly on human rights—matters deeply. In the aftermath of WWII, it was an American—Eleanor Roosevelt—who led the effort to develop a global consensus on the inherent rights and dignity of all people. Now, as then, American leadership is essential to build a world in which those universal rights are universally respected.

That is why we have focused so much of our energy and attention on getting this right. For the last decade, Human Rights First has . . . challenged arbitrary detention, torture and other cruel treatment in the wake of post-9/11 abuses. We have worked for the restoration of habeas corpus. We have served as official observers to every military commission convened at Guantánamo. And we have published a series of groundbreaking reports on Guantanamo and other aspects of U.S. detention policy. Our report, *In Pursuit of Justice: Prosecuting Terrorism Cases in the Federal Courts,* concludes that the federal system has capably handled important, complex, and challenging international terrorism cases without compromising national security or sacrificing rigorous standards of fairness and due process.

Closing Guantanamo has proven to be more difficult than many anticipated. There are tough questions to resolve, to be sure, but it is possible to close the detention facility with smart and sustained leadership from the President and Congress. Our new paper, *Guantanamo: A Comprehensive Exit Strategy*, which we are releasing today, provides a roadmap for closing Guantánamo, with practical guidance on how to address the challenges of risk management raised by transferring detainees out of the facility.

I. Why We Need to Close Guantanamo

In a world that for many is characterized by tyranny, war, and injustice, the United States stands as a beacon. Despite our many failings, the United States has a long history of advancing human rights, having played a leading role in developing the international laws and standards that define and enforce them, and continuing today by protecting refugees and supporting human rights defenders on the frontlines of the struggle for freedom in many countries around the world. Domestically, respect for freedom, democracy, and the rule of law defines our political culture and constitutional system, setting an example for people around the world who seek to advance democracy and human rights in their own societies.

A glaring exception to this narrative is the post-9/11 abuses committed by our government, defined largely by Guantanamo and the torture of detainees in U.S. custody. It's hard to overstate how much this has undermined our country's moral standing and credibility. In my role as the head of an international human rights organization, the scenarios in which I most often hear about Guantanamo are not in our domestic political debates here at home or in our courts. I hear Guantanamo raised by officials of repressive governments who use it to deflect criticism of their own policies by charging hypocrisy. And I hear about Guantanamo from human rights defenders around the world who tell me that the best thing the United States can do to support their bids for freedom and democracy is to make sure that our country can lead by example, including closing Guantanamo. Three years ago, I brought

Elisa Massimino. Committee on the Judiciary, U.S. Senate, July 24, 2013.

two dozen human rights and democracy activists from around the world to the White House to meet with President Obama, and that's exactly what they told him.

The ability of the United States to credibly push other governments to respect human rights is seriously compromised when we have failed to correct the post-9/11 abuses that have cast a shadow on America's foreign policy over the last decade. And that shadow will continue to loom large until Guantanamo is closed, and the policies of indefinite detention and military commission trials are ended.

There have been instances in the life of our relatively young country when we have pursued policies out of fear that we later realize are inconsistent with our values. Sometimes it takes hearing the perspective of those outside our national community, who know the values for which we hold ourselves out, to remind us of who we are. Consider the perspective of some family members of Guantanamo detainees. Several have written letters to you in advance of this hearing, and I want to read two excerpts from them.

Nabil Hadjarab is an Algerian man who has been detained at Guantanamo for over a decade without charge or trial. He has been unanimously cleared for transfer by our government's security and intelligence agencies. Here's what his uncle Ahmed Hadjarab wrote:

I must admit that my perception of the United States of America has been severely tarnished by this issue. When in 2002, I was told that Nabil was detained by the Americans, I thought that at least he would have a right to a fair trial. I thought his rights would be respected and that justice would prevail. What I feel today is mostly incomprehension. How can this nation, one that prides itself of defending Human Rights, close its eyes to these violations of its founding principles?

Hisham Sliti from Tunisia has been held in Guantanamo for more than a decade without charge or trial. He has also been cleared for transfer. His mother, Maherzia Sliti, wrote:

One of the worst things is the uncertainty, and the false hope that things are about to change. Sometimes I hear rumors that men have been released from Guantanamo and that Hisham is one of them. I miss and love my son so much that although my mind knows the rumors are probably false, my heart believes them every time. And every time I am devastated when I realize he is not coming home. I do not understand why my son is still in Guantanamo after all these years, when we know he has been cleared. We never thought the United States was the kind of place where people could be held like this.

These excerpts come from letters collected by Reprieve, a human rights organization that currently represents 15 prisoners in Guantanamo Bay and has provided assistance for many more. Attorneys and family members of Guantanamo detainees submitted the full versions of these and other letters to the hearing record. I encourage you to read them.

I raise these issues of justice and America's moral standing in the world because I want to be clear that what's at stake in figuring out a way to close Guantanamo is our ability to lead by example, and our reputation for upholding justice and the rule of law.

There are some who say that we need Guantanamo to hold and interrogate detainees that can't be tried in civilian court because they were captured by our military on the battlefield. But the military has never needed Guantanamo for battlefield captures; those detainees have typically been held in detention facilities in theater. Moreover, the vast majority of terrorism suspects captured abroad are dealt with by the security and law enforcement services of our foreign counterparts, and that's how it should be. Since 9/11, more than 120,000 suspected terrorists have been arrested around the world, and more than 35,000 have been convicted. Our military cannot—and should not—be the world's police force or jailor.

In cases in which we have needed to detain, interrogate, and jail terrorism suspects, our civilian system has handled these cases remarkably well. Since 9/11, civilian federal courts have handled nearly 500 cases related to international terrorism, including at least 67 where suspects were captured abroad, often in inhospitable environments. Despite claims to the contrary, there is no credible evidence that trying these cases in civilian courts has caused breaches of sensitive national security information, or invited attacks on U.S. soil.

Nor does the civilian process preclude us from obtaining actionable intelligence to disrupt terrorism plots. The administration has established a High-Value Interrogation Group (HIG) that has been deployed in a number of cases to interrogate terrorism suspects using lawful and effective methods. Even in more routine terrorism cases, and in situations where Miranda rights [such as the right no to answer question and to be represented by a lawyer] and other due process protections are respected, offering plea deals and working with the defendant's family and lawyers, in addition to lawful interrogations, have produced a wealth of actionable intelligence information, including: telephone numbers and email addresses used by al Qaeda and other terrorist groups; information about al Qaeda communications methods and security protocols; information about their recruiting and financing methods; the location of al Qaeda training camps and safe houses; information on al Qaeda weapons programs; the identities of operatives involved in past attacks; and information about future plots to attack U.S. interests.

By contrast, detention and trial at Guantanamo has proven highly problematic on several levels. Since 9/11, only 7 detainees have been convicted by military commission. Two of those convictions were recently overturned by a federal appeals court because the crimes with which the detainees were charged were not war crimes—the only acts over which military commissions have jurisdiction—at the time the offenses were committed. More broadly, in contrast to the civilian system, in Guantanamo—where detention is indefinite and Congress has made it difficult to effect transfers out of the prison—there is not the same kind of leverage (e.g., offering release or shorter detention in exchange for cooperation) to exploit with detainees.

There are other pragmatic reasons to move forward with closing Guantanamo. The impending end of combat operations in Afghanistan in 2014 increases the urgency for Congress and the administration to determine the disposition of all law-of-war detainees. The detainees at Guantanamo were apprehended and detained pursuant to the 2001 Authorization for Use of Military Force. As hostilities come to an end, Guantanamo detainees will have a legitimate claim before the courts that they should be released. Congress and the administration should proactively determine the lawful disposition of detainees now, or the courts could force those dispositions later.

There has long been a national security consensus that Guantanamo should be closed. More than 50 retired generals and admirals, along with three Secretaries of Defense—[Robert] Gates, [Leon] Panetta and [Chuck] Hagel—have called for Guantanamo to be closed. Today's witnesses underscore that many senior military leaders believe that closing Guantanamo is a national security imperative.

As a national security issue, closing Guantanamo should be beyond politics. And it has been in the past. In 2008, there was significant bipartisan consensus that Guantanamo should be closed. Then-President [George W.] Bush said he wanted to close Guantanamo, as did then-candidates [Barack] Obama and [John] McCain. That consensus has started to re-emerge, with Senator McCain recently stating that Guantanamo should be

closed, and emphasizing that it would be an "act of courage" to transfer detainees out of Guantanamo and into the United States as part of a plan to close the facility.

And Guantanamo can be closed—safely and securely. This is not to say that closing Guantanamo will be easy—if it were, Guantanamo would already be closed. There are difficult legal, practical, and political problems that must be addressed to move forward.

But there is a pragmatic path forward to close Guantanamo, if the administration and Congress demonstrate sustained and focused leadership to get the job done.

I want to spend a few minutes outlining this path forward.

II. A Comprehensive Plan for Closing Guantanamo

In 2009, President Obama signed an executive order establishing an interagency taskforce to conduct a review and recommend lawful dispositions of the detainees being held at Guantanamo. Since then, 72 prisoners have been transferred, repatriated or resettled, and a number of other detainees have died—either by suicide or other causes—bringing the current detainee population down to 166. Transfers have stalled in part because of restrictions imposed by Congress in 2010, 2011 and 2012, and because the administration has failed to exercise the authority Congress gave it in 2012 under the National Defense Authorization Act to waive the transfer restrictions by invoking, among other requirements, national security interests.

Concerns about recidivism—the possibility that a released detainee may "return to the fight"—are understandable, as they are in the criminal context. But, as many analysts have detailed, the claims about recidivism of detainees who have already been released are inflated. The claim by members of Congress and some in the media that 28% of former Guantanamo detainees have "rejoined the fight" or "returned to the battlefield" is highly misleading. It appears to be based on unreliable or unconfirmed reports of suspected activities, and in any event includes detainees that may not have participated in any terrorist plots or attacks. The process to evaluate potential transfers has changed since the prior administration to more accurately capture post-transfer risk, leading to fewer cases of recidivism for detainees transferred by the Obama administration. The Director of National Intelligence's recidivism assessment should be revised to more accurately reflect the circumstances in which former detainees that have engaged in terrorist plots or attacks against the United States so that evaluation—and mitigation—of this risk is grounded in reality, not hyperbole.

Nonetheless, as senior military commanders have told me, transfers of detainees from Guantanamo—just as transfers of detainees out of detention facilities in Iraq and Afghanistan—have always been about risk management, not risk elimination. Some detainees pose little risk; others will pose more. Establishing a "zero tolerance for risk" policy with respect to individual detainees is neither wise nor necessary. Our military, intelligence, law enforcement, and diplomatic agencies, along with those of our foreign counterparts, can significantly mitigate the risks of transferring detainees out of Guantanamo through security assurances, monitoring, rehabilitation and other reasonable measures. The risks associated with keeping Guantanamo open are harder to mitigate, and the harm is potentially far more lasting.

A. Disposition of the 86 Detainees Cleared for Transfer

Of the 166 detainees remaining at Guantanamo, 86 have already been cleared by all relevant law enforcement, defense, and intelligence agencies for transfer back home or to third countries. The United States has determined that those men should neither face trial nor be detained, and many were cleared for transfer by both the Bush and Obama administrations. Several of these men have languished at Guantanamo for more than eleven years, even as their home countries have demanded their return. To successfully transfer all or most of these 86 detainees, the administration should take the following steps.

The Secretary of Defense should certify transfers and issue national security waivers to the fullest extent possible consistent with applicable law. The current set of certification requirements, coupled with the national security waiver, provides the administration with the authority to transfer many, if not all, of the 86 detainees who have already been cleared for transfer. In most cases, security assurances from, or changes in the political or security context in, the receiving country can be read to satisfy the remaining certification requirements that cannot be waived under the national security waiver. Efforts to negotiate any required assurances should begin immediately and be given the highest priority under the direction of the Secretary of Defense, in concurrence with the Secretary of State and in consultation with the Director of National Intelligence, pursuant to the required statutory guidelines.

The administration should begin transferring individuals to Yemen on a case-by-case basis, while also expeditiously developing a rehabilitation program there that could facilitate transfers of cleared Yemeni detainees en bloc. Fifty-six of the 86 detainees cleared for transfer are from Yemen. Of those 56, 26 are cleared for transfer without conditions, and may be transferred now that the moratorium on transfers to Yemen has been lifted. The remaining 30 are conditionally cleared for transfer, and may be transferred with improved security conditions in Yemen, an appropriate rehabilitation program, or where third-country resettlement becomes an option.

The administration should transfer home the 17 cleared detainees who are from countries that have requested their return (other than Yemen, which has also requested its citizens back). Countries that have demanded the return of their cleared citizens include: Afghanistan (4), Algeria (5), Libya (1), Saudi Arabia (2), and Tunisia (5). However, in accordance with U.S. non-refoulement obligations, where there are substantial grounds for believing that a detainee would be in danger of being subjected to torture or other forms of mistreatment if returned home, the administration should resettle such detainees in third countries. [Non-refoulement is a standard of international law that forbids handing over someone who is being persecuted (not simply prosecuted) to the persecutor(s).]

The administration should transfer to third countries (including, possibly, the United States) the three Uighur detainees who cannot be repatriated to China based on their well-founded fear of persecution. [Uighurs are an ethnically Turkic, mostly Muslim people living in northwestern China.] The Uighurs are not part of al Qaeda, the Taliban or any "associated forces," and do not pose a material threat to the United States. U.S. federal courts have ordered their release. Moreover, resettling such detainees here would place the United States in a stronger position to negotiate transfers of other detainees to third countries by demonstrating a willingness to share in the responsibility of resettlement.

The administration should transfer home the five men whose countries have not, at least publicly, asked for their citizens back, including men from Mauritania, Morocco, the Palestinian Territories, Tajikistan, and the United Arab Emirates. These detainees should be repatriated home if their countries are willing to accept them and transfers can be effectuated consistent with non-refoulement obligations. If that is not possible, they should be resettled in third countries. Finally, the administration should transfer to third countries the four Syrian detainees and one Sudanese detainee who cannot be repatriated because federal law prohibits transfers to Syria and Sudan as state sponsors of terrorism.

Congress has a role to play in facilitating the responsible closure of Guantanamo. The annual defense bill reported out of the Senate Armed Services Committee presents the opportunity for a compromise approach on the resettlement or repatriation of detainees. While it requires the Secretary of Defense to take steps to mitigate the risks of transfer and to consult with Congress about decisions made, it properly places such decisions with the defense and intelligence agencies that are better situated than Congress to make those decisions. This legislation places unnecessary restrictions on the President's ability to close Guantanamo, but it is certainly an improvement on the current absolute bar on transferring detainees to the U.S.

B. Disposition of the 34 Detainees Suspected of Criminal Conduct and Slated for Prosecution

The Guantanamo Review Task Force designated 34 of the remaining 166 detainees at Guantanamo as eligible for prosecution before either a federal court or military commission. Recent federal appellate court decisions have overturned two military commission convictions because the crimes for which the detainees were convicted—material support and conspiracy—were not internationally-recognized war crimes at the time of the offense.

As a result, there may now be only twenty men who could face trial by military commission, though they and other detainees in this category could possibly face prosecution in an Article III [of the U.S. Constitution] federal court should Congress permit transfers to the United States for prosecution. In addition, the current 9/11 cases and the case of the alleged USS Cole bomber before military commissions have been beset with scandal (the CIA was discovered to have the ability to censor the proceedings) and legal uncertainty (the presiding judge could not even rule whether the constitution applies). [The courts in the central three-tier system of federal courts (district courts, the circuit courts of appeals, and the U.S. Supreme Court) are all established under Article III of the Constitution. Military commissions and other military courts marshal and some other federal courts are Article II courts, created under the powers of Congress.]

In order to resolve the cases of the 34 detainees in this category, Congress should pass the Senate version of the National Defense Authorization Act (S. 1197), reported out of the Armed Services Committee, which removes the ban on use of Pentagon funds for transfers to the United

States for prosecution, incarceration and medical treatment. The administration cannot currently issue national security waivers to ensure prosecution of these detainees; Congress must act. If transfers to the United States are again allowed, the administration should transfer those already facing military commissions at Guantanamo to military commission trials in the United States in order to facilitate the closing of Guantanamo. Military commissions should be used only to resolve the legacy cases at Guantanamo, not to supplant Article III federal courts, which have proven more muscular and adept in counterterrorism prosecutions. Article III courts have four times the number of substantive criminal laws available to them for use against terrorism suspects—not to mention more than two hundred years of experience and precedent on which to rely.

The administration should therefore transfer any remaining detainees who can be charged with crimes to a civilian court in the United States, or to an appropriate foreign jurisdiction, where such transfers can be made consistent with applicable law. Ahmed Ghailani, a former Guantanamo detainee, was transferred to the Southern District of New York, convicted, and is now serving a life sentence.

The administration should transfer those already convicted to any appropriate high security federal prison, which can safely house detainees. There are 355 terrorism convicts serving sentences in United States federal prisons, including the only 9/11 defendant to stand trial—Zacarias Moussaoui—who was convicted in federal court and sentenced to life in prison as the alleged 20th hijacker on 9/11. Three Guantanamo detainees have already been convicted by military commission and are serving sentences at Guantanamo. Those who suggest that detainees who have served their sentence would be set loose on the streets of America are misinformed. Any such person would be subject to mandatory deportation.

C. Disposition of the 46 Detainees That Have Neither Been Charged with a Crime Nor Been Cleared for Transfer

The remaining 46 out of 166 detainees being held at Guantanamo will either have to be charged with a crime or, eventually, be released within some reasonable period of time at the end of combat operations in Afghanistan or some other appropriate marker of the end of hostilities. That is what is required under the laws of war and our Constitution, and it is what we have done at the end of past conflicts. The United States transferred 10,000 prisoners to the Iraqi government at the end of the Iraq war, and has already transferred control of thousands of detainees in Afghanistan to that country's government. Finding a reasonable, lawful disposition for this group may be more challenging, but it is not insurmountable.

First, the administration must initiate the Periodic Review Board [PRB] hearings pursuant to Executive Order 13567 for eligible detainees immediately under the direction of the Secretary of Defense. No congressional action is needed to do this. These hearings should be completed in a timely and effective manner to determine whether each detainee is eligible for transfer. In an encouraging development, the Pentagon has announced that PRB hearings will begin soon, though it has not said when, or established a timeline for completing the hearings.

The Periodic Review Boards should, consistent with the interests of national security, afford detainees access to evidence, counsel and other markers of due process to ensure a thorough and accurate review. The boards could determine that some number of men in this group is now eligible to be transferred because new evidence has surfaced, the political situation in their country has improved, their networks of influence have degraded, their health has deteriorated, or other factors, such that they no longer pose a significant risk.

The administration should also provide Periodic Review Board hearings for any detainees who were previously slated for prosecution whom the administration no longer intends to prosecute. Timely and effective hearings should determine whether continued detention is necessary to protect against a significant threat to the security of the United States.

The administration should also determine whether there are extant credible criminal charges in other foreign jurisdictions where the detainees could be tried.

The administration should determine whether the 10 Afghan detainees of the 46 held in this indefinite detention category can be repatriated to Afghanistan pursuant to a negotiated agreement with the Taliban or the government of Afghanistan. Likewise, the administration should determine whether the 26 Yemeni men held in this category can be transferred based on coordination with the state of Yemen. Long-term efforts by Yemen to institute a rehabilitation program could assist in the transition.

Lastly, the administration could transfer some number of Guantanamo detainees to the United States for continued detention or trial until the end of hostilities. Some have expressed concerns that doing so could embed the injustices of Guantanamo's indefinite detention scheme in domestic practice. While we don't discount those concerns, we believe that with appropriate safeguards to ensure against the use of indefinite detention and military trial authorities for future captures, transfer of detainees to the United States is an acceptable option in furtherance of a broader effort to close Guantanamo. The Government Accountability Office [GAO, an investigative agency of Congress] has documented the high security prison facilities in the United States with capacity that could hold detainees.

To the extent that the administration has not resolved the disposition of any detainees prior to the end of hostilities, the administration should repatriate or resettle these detainees at the end of combat operations in Afghanistan or some other reasonable marker of the end of hostilities.

III. Conclusion

In one sense, closing Guantanamo is a numbers problem—how to get from 166 to zero. Once there were 779 prisoners at Guantanamo. The Bush administration resettled or repatriated more than 500 of them. The Obama Administration has gotten that number down to 166, a majority of whom have been cleared for transfer by the Department of Justice, Department of Defense, Department of State, Department of Homeland Security, Office of the Director of National Intelligence, and Joint Chiefs of Staff. The remaining task is about managing risk to achieve an important national security objective on which there is bipartisan consensus. The risks of transfer can be mitigated; the risks of maintaining Guantanamo forever cannot.

But in another sense, closing Guantanamo is about who we are as a Nation. As the President recently said:

> I know the politics are hard. But history will cast a harsh judgment on this aspect of our fight against terrorism, and those of us who fail to end it. Imagine a future—ten years from now, or twenty years from now—when the United States of America is still holding people who have been charged with no crime on a piece of land that is not a part of our country. Look at the current situation, where we are force-feeding detainees who are holding a hunger strike. Is that who we are?

At a certain point, who we are as a nation cannot be separated from what we do. Guantanamo is a symbol for many around the world of torture, injustice and illegitimacy. As the United States winds down the war in Afghanistan, Congress and the President have the opportunity to transform this legacy and restore America's reputation for justice and the rule of law.

I urge you to align our actions with our ideals and work with the President to get this done.

ELISA MASSIMINO is president and chief executive officer of Human Rights First, a U.S. based advocacy organizations seeking to advance U.S. global leadership on human rights. She holds a law degree from the University of Michigan.

Frank J. Gaffney, Jr.

 **NO**

Testimony During Hearings on "Closing Guantanamo: The National Security, Fiscal, and Human Rights Implications"

As a former member of the staff of a great Democratic Senator, Henry "Scoop" Jackson [D-WA], and as a professional staff member for the Senate Armed Services Committee under Republican Chairman John Tower [TX], I have great affection for this institution. I revere the mandate it received from the founders as a co-equal partner with the executive in governing this nation.

In my subsequent four-and-a-half years in the [President Ronald] Reagan Defense Department—in which, among other capacities, I acted as the Assistant Secretary of Defense for International Security Policy, I had a different perspective on the accountability the Congress could exact from the executive branch. But I welcomed then, and encourage now, the legislature's indispensable oversight role—a role that is, in my view, essential to maintaining a "well-ordered liberty."

The Case for Gitmo

Let me begin my argument for retaining the detention and interrogation facility at Guantanamo Bay, Cuba (nicknamed "Gitmo") by noting a fundamental reality: **Our nation is at war** [emphasis in the original]. We are operating in that status pursuant to Congress's 2001 Authorization for the Use of Military Force (AUMF), and in accordance with the laws of armed conflict governing a nation's right to self-defense. These are the legal mechanisms of which we have availed ourselves to enable and guide the use of force necessary to protect the United States.

We have been obliged to go to war because it was thrust upon us. And, if we are to prevail in this conflict, we must understand the nature of the enemies with whom we are at war. They are *shariah*-adherent jihadists who believe, in accordance with that doctrine, that it is God's will that they destroy our way of life and subjugate us to theirs. [The shariah (or sharia) is Islamic religious law.]

It is important to state at this point that not all Muslims subscribe to shariah, or seek to impose it on the rest of us. Those that do not adhere to this ideology are not necessarily a problem. They could even be critical to mitigating the threat posed by their co-religionists who *do* embrace shariah. But it is a grievous mistake to think that those we confront are not animated by what they believe to be a spiritual mandate, that we confront only threats from al Qaeda, or that its members are appreciably distinct from others who pursue shariah's requirements to achieve its supremacy worldwide under the rule of a caliphate [an Islamic country ruled by a religious leader under shariah].

Our shariah supremacist enemies have made their intentions known to us prior to the devastating attacks on 9/11, and they have made no secret of them since. The belief that their holy war is divinely inspired has contributed not only to the violent and stealthy forms of jihad being waged against us. It has also contributed materially to the determination of a significant percentage of those captured on the battlefield and detained at Guantanamo Bay to return to the fight if and when they are released.

It would be the subject for another, most useful hearing if this Committee were to examine the lengths to which we have gone as a nation to ignore these realities. Suffice it to say for the present purpose that, by failing to understand the nature and abiding ambition of our foes, we are prone to making dangerous tactical decisions, such as releasing hardened detainees, and potentially fatal strategic ones, including contemplating the closure of Gitmo.

Let's be clear: **Guantanamo Bay is the optimal location for U.S. detention and interrogation of unlawful enemy combatants** [emphasis in the original]. It is simultaneously a uniquely secure facility and a highly humane one. And Gitmo has these attributes primarily thanks to the servicemen and women whose professionalism, discipline and courage make them possible notwithstanding routine, vile and often violent provocations on the part of detainees.

The Absence of Sound Alternatives

The burden of proof should be on opponents of Gitmo to define a superior arrangement. To date, they have been unable to persuade the Congress that there is such an alternative. Indeed, the other choices pose grave risks for national security and/or are less humane than incarceration at Guantanamo Bay. Let me briefly examine several of these in turn.

First, handing detainees over to third-party nations can result in the prisoners deliberately being set free, breaking out of jail or otherwise being enabled to re-join fellow jihadists on the battlefield [emphasis in the original]. In 2010, the Obama administration suspended the transfer of detainees to Yemen out of concern that, according to the *Washington Post*, "a deteriorating security situation driven by a branch of al-Qaeda has stoked fears that detainees could join—or rejoin—the terrorist organization if released."

Just yesterday, the Iraqi arm of al Qaeda claimed responsibility for raids on prison facilities near Baghdad that released hundreds of inmates, including members of al Qaeda. This incident shows the folly of relying on vulnerable foreign prisons to keep dangerous individuals incarcerated. The risk of former Guantanamo Bay detainees returning to the battlefield is a significant one.

Last year, the Office of the Director of National Intelligence [ODNI] released a report indicating that, of the 599 released former Gitmo detainees, 27.9% were either confirmed or suspected of engaging in terrorist activity. This amounts to a 2.9% increase in former Guantanamo detainee recidivism as reported by the ODNI in December 2010. My guess is that some number of the remaining group is *also* back in the jihad, even if there is no evidence of it thus far. **Second, transferring the Guantanamo detainees to the United States for detention—in say a prison like that formerly known as the Thomson Correctional Facility in Illinois—poses substantial security risks** [emphasis in the original]. For one thing, there is the danger arising from what the jihadi detainees might do inside a U.S. prison population in terms of violent plots or perhaps simply their toxic form of shariah proselytization.

For another, housing prominent jihadists in a given American community could cause it to be targeted by their comrades, either in the hope of actually freeing the detainees or simply as an act of jihad. Former federal prosecutor Andrew C. McCarthy, who secured the conviction of the "Blind Sheikh" [Sheikh Omar Abdel-Rahman, an Egyptian living in the United States] for his role in the first bombing of the World Trade Center in 1993, has previously pointed out that jihadists target military bases, and U.S. military bases consist of entire communities where members of our Armed Forces live with their families.

Once detainees are physically inside the United States, moreover, they are within the jurisdiction of federal judges, before whom defense attorneys will argue their clients deserve the full array of constitutional rights afforded to common criminals. Undoubtedly, some federal judges will agree with this assertion.

That would, in turn, enable detainees to be tried in this country under criminal law standards that cannot, as a practical matter, be applied to the circumstances of wartime capture (e.g., evidentiary procedures, Miranda rights, etc.) Prosecutors could then be put in the position of having to disclose classified information in order to secure a conviction under these standards, or risk having the detainee be released—perhaps *inside the United States* [emphasis in the original] especially if no other country is willing to take him.

Let's not kid ourselves. Even if such risks were non-existent, or simply deemed acceptable, there is no reason to believe that holding Gitmo detainees would spare us the criticism of human rights advocates and defense lawyers of the "Gitmo bar"—including appointees in the Obama administration. As Andy McCarthy has also noted, some of these folks have previously asserted that Supermax-type confinement is a human rights violation. In point of fact, "shoe-bomber" Richard Reid, who was held in a Supermax facility under "special administrative measures" (SAMs) to ensure his secure confinement, argued that the SAMs violated his constitutional rights. The SAMs were subsequently lifted.

Finally, it has been asserted that the existence of Guantanamo Bay has served as a "recruitment tool" for terrorists and that the facility should be shut down for that reason. In fact, shutting down Guantanamo Bay detention operations would rightly be seen by the jihadist movement worldwide as evidence of our submission, and a greatly emboldening victory [emphasis in the original]. It would likely have the effect of increasing recruitment, while at the same time denying us a vital tool for incarcerating and interrogating those we capture rather than kill.

What is more, such a victory would embolden not only the violent jihadists, but also the pre-violent jihadists (most prominently the Muslim Brotherhood), here and abroad. The latter seek the same outcome as the former—the imposition globally of shariah under the rule of a new caliphate. The only difference is one of tactics driven by the Brotherhood's perception that, for the moment, the correlation of forces is not conducive to success via direct and violent forms of jihad.

Conclusion

For all of these reasons, it is, in my professional judgment, not only desirable but necessary to continue to incarcerate detainees at Guantanamo Bay. We should, moreover,

be free to add to their number at Gitmo, if that will help us gather vital intelligence and keep dangerous jihadist enemy combatants off the battlefield.

Frank J. Gaffney, Jr., is founder and president of the Center for Security Policy, a private analysis and advocacy organization. He holds an MA degree from Johns Hopkins University's Paul H. Nitze School of Advanced International Studies.

EXPLORING THE ISSUE

Is Closing the Prison for Alleged Terrorists at the U.S. Naval Base, Guantanamo, Cuba Justified?

Critical Thinking and Reflection

1. Would you give all the constitutional rights of people accused of a crime to all the prisoners in the Guantanamo Bay detention facility? If not, which ones would you deny them? To think about this, find a list of such rights in the criminal justice system. One brief source is *Your Basic Constitutional Rights in the Criminal Justice System*, authored by the Carl Vinson Institute of Government, University of Georgia and located on the web at www.georgialegalaid.org/resource/your-basic-constitutional-rights-in-the-crimi?ref=e31UD.
2. How would you deal with the argument that in some cases producing the evidence, particularly in an open trial, against some terrorists suspects would jeopardize U.S. intelligence agents, damage the ability to conduct covert intelligence, and even put U.S. intelligence agents, informants, witnesses, and others in grave danger?
3. Would it be better to have foreign terrorists apprehended abroad tried in an international court than in an American court?

Is There Common Ground?

Common ground is possible, but getting to it will be difficult. Most of those ever held at Gitmo have been released because they had little or no involvement in terrorism. Stepped up efforts can be made to "place" the 75 or so prisoners cleared for release to another country. The difficult part is the fate of the 40 or so prisoners who the U.S. government deems too dangerous to release. They can be tried and imprisoned under the military commission system, but bringing them back to the United States to stand trial or even to be detained is not possible under laws passed by Congress. Moreover, there is strong public resistance to any such move. A 2014 Gallup poll revealed that 66 percent of Americans opposed bringing any of the prisoners into the country for trial or incarceration. Only 29 percent supported such a move. Furthermore, that approximate division of opinion has been consistent over time. According to a *Huffington Post* poll in 2012, a solid majority (63%) Americans favor trials for the detainees, but a majority (52%) favored trials conducted by military commission, while only on 28 percent supported trials in U.S. civilian courts.

Create Central

www.mhhe.com/createcentral

Additional Resources

Bazelon, Emily. (June 8, 2014). *Gitmo Fail. Slate* online at: www.slate.com/articles/news_and_politics/politics/2014/06/bowe_bergdahl_prisoner_swap_obama_finally_releases_detainees_from_gitmo.html

A recent review of the attitudes and actions of the Obama administration related to Gitmo. It begins with the statement and question: "Obama promised to close Guantánamo. Why is he releasing dangerous detainees and ignoring the rest?"

Bravin, Jess. (2013). *The Terror Courts: Rough Justice at Guantanamo Bay* (Yale University Press)

Ellis, T. S. III. (2013). "National Security Trials: A Judge's Perspective." *Virginia Law Review* (99:1607–1652)

Thomas, Donnie L. (2013). *Joint Task Force–Guantanamo Bay, Cuba: Open or Close?* (Army War College Carlisle Barracks)

Internet References . . .

Guantanamo

Maintained by Al Jazeera America (AJAM), an America-based cable and satellite news television channel, vaguely akin to CNN, that is a subdivision of Al Jazeera, the Arab/Middle East/Muslim media outlet based in Qatar and owned by its royal family.

http://america.aljazeera.com/topics/topic /international-location/americas/guantanamo -bay.html

Guantánamo by the Numbers—The American Civil Liberties Union

Data on prisoners, costs, and other factors compiled by the American Civil Liberties Union.

https://www.aclu.org/national-security /guantanamo-numbers

Guantanamo Docket: Detainees.

A *New York Times* site that lists all 779 individuals imprisoned since 2002 at Guantanamo, their nationalities, and their status. Hyperlinks with each name give some information about the individuals.

http://projects.nytimes.com/guantanamo/detainees

U.S. Department of Defense, Guantanamo Bay

Two sites with details from the U.S. government point of view, with particular information on the detainees and the charges and proceedings against them.

http://www.defense.gov/home/features/detainee _affairs/

http://www.defense.gov/home/features/gitmo/

Unit 5

UNIT

International Law and Organization Issues

*W*hile international relations is nearly as old as human political governance, the frequency, scope, and importance of interactions between and among countries has grown with increasing speed, especially during the past two hundred years or so and even more extraordinarily since World War II. One factor is technological advances. Technology has increased our ability to extract resources and produce goods, and that has expanded our need to get resources and sell our products to others through trade. Transportation advances have also advanced trade. Modern communications and transportation have rapidly expanded the speed with which merchandise, money, people, information, and ideas move over long distances.

This growing interchange and other factors have led to problems becoming more and more international and less and less purely national in nature. Technology is responsible for global warming, a planetary problem. Once little-known, seemingly faraway diseases like Ebola are now rightly seen as global threats. Terrorism ignores borders, and nuclear war could extinguish countries and their borders altogether. Economic downturns now happen to the world as a whole, with the global mega-recession beginning in 2008 the most recent example.

More aspects of the coming together of the world for good or ill could be added, but what is important for this unit on international law and organization is that both have also expanded as part of an effort to regulate the much higher levels of international interaction and the ever more international nature of the major problems people and their countries face. There has been a vast growth of international organizations since 1900. Some are general like the UN; others are regional like the European Union. Some are general, again like the UN; others are specialized, like the World Trade Organization. None existed 125 years ago. Similarly, the need for the world to act predictably has led to a burgeoning number of multilateral treaties that govern the behavior of countries in an expanding range of matters governed by international law. Because international law is shackled by limited enforcement mechanisms, it is far less effective than domestic law. Still, there is far more international law now than there was not long ago and far less than there will be in the future.

ISSUE

Selected, Edited, and with Issue Framing Material by:
John T. Rourke, *University of Connecticut, Storrs*

Is the UN a Worthwhile Organization?

YES: Susan E. Rice, from "Six Reasons the United Nations Is Indispensable," address delivered at the World Affairs Council of Oregon, Portland, Oregon (2011)

NO: Bruce S. Thornton, from "The U.N.: So Bad It's Almost Beautiful," *Hoover Digest* (2012)

Learning Outcomes

After reading this issue, you will be able to:

- Relate some of the contributions that the United Nations (UN) has made over the years and some of the charges against the UN by critrics.
- Explain why reform is so difficult at the United Nations.
- Discuss each of the six reasons that make the UN indispensable according to Ambassador Rice.
- Comment on whether from the U.S. perspective it would be better to try to reform the UN even more or simply withdraw from it.

ISSUE SUMMARY

YES: Susan E. Rice, U.S. ambassador to the United Nations, tells an audience that the United States is much better off—much stronger, much safer, and more secure—in a world with the United Nations than the United States would be in a world without the UN.

NO: Bruce S. Thornton, a research fellow at the Hoover Institution at Stanford University in California, writes that the United Nations is fatally flawed by not having consistent, unifying moral and political principles shared by member nations that can justify UN policies or legitimize the use of force to deter and punish aggression.

The United Nations was established in 1945 as a reaction to the failure of its predecessor, the League of Nations, to prevent World War II and its horrendous destruction of life and property. From this perspective, founding the UN in 1945 represented something akin to a sinner resolving to reform. In this case, a world that had just barely survived a ghastly experience pledged to organize itself to preserve the peace and improve humanity. Now, nearly 70 years later, international violence continues, global justice and respect for international law remain goals rather than reality, and grievous economic and social ills still afflict the world.

It is also the case that the UN has been and remains principally a political organization in which the countries of the world maneuver to advance their political agendas rather than an international organization in which the world's countries work together in the spirit of cooperation to solve global problems. Moving forward is often particularly difficult because most peacekeeping activity using UN forces and many other key UN functions need the approval of the Security Council. But in a scheme that dates to the power realities in 1945 when the UN was organized, the Council is dominated by five permanent members, each of which can veto almost anything the Council does. These members are China, France, Great Britain, Russia, and the United States. In recent times, for example, efforts to put strong UN-backed sanctions on Iran in response to its alleged nuclear weapons program have been blocked in the Security Council by the unwillingness of China and Russia to support tough sanctions.

The United States also often finds itself outvoted in the other main UN legislative body, the General Assembly. For a quarter century most UN members were U.S. allies, and Washington was usually able to dominate the General Assembly, where each country has one vote. However, the UN's membership changed during the 1960s and 1970s when dozens of former colonies in Africa, Asia, and elsewhere gained independence and joined the UN. These new member countries often saw things differently than the United States, and by the 1980s the United States' ability to almost always muster a majority in the UN General Assembly had waned.

There have also been widespread charges that the UN's bureaucracy is excessive and inefficient. Partly in reaction to that, Congress in 2002 mandated a 12 percent reduction in the percentage of the basic UN budget funded by the United States and has rarely appropriated enough money to even meet the new, lower figure. Even more recently, several scandals have rocked the UN. From 1996 to 2003, it ran an "oil-for-food" program that permitted Iraq to sell oil and use the receipts to buy food, medicine, and other humanitarian supplies under UN supervision. Persistent rumors about corruption in the $67 billion program led to an investigation in 2004 that found substantial corruption, including the program administrator

taking brides to ignore the fact that Iraq was diverting huge sums to the purchase of munitions and for other banned uses. The UN's image was further sullied in 2004 when evidence came to light of UN peacekeeping troops and personnel trading food and other necessities of life to obtain sex from those they were supposedly protecting.

Yet for all the flaws and frustrations, the UN has made many contributions. Since the UN was founded, it has fielded 67 peacekeeping operations using hundreds of thousands of military personnel, tens of thousands of UN police offices, and many support personnel from 120 countries. More than 3,000 of these peacekeepers have died while serving under the UN flag. The UN has also helped shelter and fed millions of refugees, has supplied agricultural and other developmental assistance to more than half the world's countries, and has made many other contributions. Are these enough to offset the UN's limits and flaws? In the YES selection, Susan Rice answers yes. She contends that while the UN is not perfect, it plays an indispensable role in promoting U.S. interests and values. Bruce Thornton disagrees in the NO selection. He criticizes Americans who believe in the usefulness of the UN. He contends that the UN is a collection of unaccountable functionaries of tyrannous regimes—regimes that undermine the legitimacy of UN policy.

YES

Susan E. Rice

Six Reasons the United Nations Is Indispensable

We're meeting at the end of an extraordinary day—a rare moment in our lives when we have had the privilege to witness history in the making. Egyptians have just overthrown long-time, authoritarian president Hosni Mubarak. The proud people of Egypt have reminded the world of the power of human dignity and the universal longing for liberty. The American people have been deeply inspired by the scenes in Cairo and across that great ancient land. President Obama today recalled the words of Dr. Martin Luther King Jr., who said, "There is something in the soul that cries out for freedom." As those cries came from Tahrir Square [in Cairo], they moved the entire world. The United States will fully support a credible and irreversible transition to genuine democracy in Egypt.

February 11 [2011] is turning out to be one of those dates that echoes in history. How many of you remember that 21 years ago today, Nelson Mandela walked out of prison? [Mandela was a leader of the effort to end the white-dominated racist government of South Africa. He became the country's first president under the new, reformed government.] There were those who said that South Africa could not handle democracy, that it would set off a wave of instability and violence. Instead it set off a wave of liberty for South Africans. Many people thought South Africa couldn't do it. South Africa proved the naysayers wrong. I'm very confident the people of Egypt can do the same. Now, let me turn from a day of astonishing change to some of the changes we have made in America's approach to the wider world.

We're now two years into the Obama Administration. At a time of economic trial and sweeping change, we've made America stronger and more secure by pursuing a strategy of national renewal and energetic global leadership. Tonight, I want to discuss how the United Nations fits into that strategy—why we need the UN, how it makes us all safer, and what we're doing to fix its shortcomings and help fulfill its potential.

In these tough economic times, we're focused on getting our economy growing and providing jobs to Americans who're hurting. Yet even as we get our own house in order, we cannot afford to ignore problems beyond our borders. When nuclear weapons materials remain unsecured in many countries around the world, all our children are at risk. When states are wracked by conflict or ravaged by poverty, they can incubate threats that spread across borders—from terrorism to pandemic disease, from criminal networks to environmental degradation. Like it or not, we live in a new era of challenges that cross borders as freely as a storm—challenges that even the world's most powerful country often cannot tackle alone. In the 21st century, indifference is not an option. It's not just immoral. It's dangerous.

Now more than ever, Americans' security and wellbeing are inextricably linked to those of people everywhere. Now more than ever, we need common responses to global problems. And that is why America is so much better off—so much stronger, so much safer and more secure—in a world with the United Nations than we would be in a world without it.

Main Street America needs the United Nations, and so do you and I, especially in these tough economic times. America can't police every conflict, end every crisis, and shelter every refugee. The UN provides a real return on our tax dollars by bringing 192 countries together to share the cost of providing stability, vital aid, and hope in the world's most broken places. Because of the UN, the world doesn't look to America to solve every problem alone. And the UN offers our troops in places like Afghanistan the international legitimacy and support that comes only from a Security Council mandate—which, in turn, is a force multiplier for our soldiers on the frontlines. It is all too easy to find cases where the UN could be more efficient and effective. I spend plenty of time pointing them out and trying to fix them—and not always diplomatically. But judging the UN solely by isolated cases of mismanagement or corruption misses the forest for the trees. We're far better off working to strengthen the UN than trying to starve it—and then having to choose between filling the void ourselves or leaving real threats untended.

Rice, Susan E. From remarks delivered at World Affairs Council of Oregon, Portland, OR, February 11, 2011.

The American public—you get it. An October 2010 survey . . . found that 72 percent of Americans support paying our UN peacekeeping dues in full and on time. The American people are fundamentally pragmatic. They know, after all, that America created the UN. In 1942, during World War II, Franklin Delano Roosevelt summoned 26 allies to Washington to sign the Declaration of the United Nations and pledged "to defend life, liberty, independence and religious freedom." As President Truman subsequently boasted, "We started the United Nations. It was our idea." Years later, President Reagan affirmed: "We are determined that the United Nations shall succeed and serve the cause of peace for humankind.

Roosevelt and Truman were practical men who wanted common action to halt aggression and prevent another world war. As one of the UN's greatest Secretaries-General, Dag Hammarskjöld [in office, 1953–1961], put it, the UN was designed "not to bring humanity to heaven but to save it from hell." Over the years, we've learned the price of letting global problems go unaddressed. So the UN has taken on huge responsibilities for keeping the peace—and for saving innocent civilians not just from the hell of conflict but also from the hell of displacement, disease, and despair.

The truth is: the UN has also picked up some bad habits along the way, and we must continue to be clear about its shortcomings. You hear a lot of criticism of the UN from some quarters—and, I confess, I agree with some of it. But we must not lose sight of the many burdens the UN helps shoulder and the many benefits it provides to every American.

Some critics argue that we should withhold our UN dues to try to force certain reforms, or that we should just pay for those UN programs we like the most. This is short-sighted, and it plain just doesn't work. The United States tried this tactic during parts of the 1980s and 1990s, and the result was that we were more isolated and less potent. That is because great and proud nations like ours are judged by their example. They are expected to keep their treaty commitments and pay their bills. When we shirk our responsibilities, our influence wanes, and our standing is diminished. Imagine going to a restaurant, getting a pretty good steak that could have been cooked a little better, and then skipping out on the check. We just cannot lead from a position of strength when we're awash in unpaid bills. We cannot depend on UN missions in Iraq and Afghanistan to help our troops return home safely and in success—and then decimate the budgets that fund them. And, if we treat our legally binding financial obligations like some kind of a la carte menu, we invite others to do the same. So, instead of paying just 22 percent of the nearly half a billion dollar annual cost of crucial UN support operations in Iraq and Afghanistan, we'd be stuck with almost the whole tab ourselves.

Yet paying our bills in full and on time doesn't mean giving the UN a pass. As we work with Congress in a bipartisan spirit to meet our responsibilities, we continue to lead the charge for serious and comprehensive reform. The UN has far more to do to create a culture of economy, ethics, and excellence. The UN must be more lean, more nimble, and more cost-effective. In recent years, the United States has spurred important changes, including revitalizing the UN Ethics Office, now headed by a respected American. The newly appointed UN inspector-general is a tough Canadian auditor committed to whipping into shape an atrophied investigations division. And no one has pushed harder than the U.S. to protect whistleblowers, impose budget discipline, and promote transparency. But the UN still has much to do to reduce bureaucracy, reap savings, reward talent, and retire underperformers.

Some Americans believe the UN infringes on American sovereignty. Frankly, I am baffled by this concern. The fact is: the UN can't tax us. It can't override U.S. law. The UN can't order our soldiers into battle. It can't take away our Second Amendment rights. The UN can't impose social norms on us. And it doesn't begin to have any much-hyped fleet of secret black helicopters. The truth is: the UN Security Council can't even issue a press release without America's blessing. The UN depends entirely on its member states, not the other way around. When the UN stumbles, it's usually because its members stumble—because big powers duck tough issues in the Security Council or spoilers grandstand in the General Assembly. As one of my predecessors, the late Richard Holbrooke, was fond of saying, "Blaming the UN when things go wrong is like blaming Madison Square Garden when the Knicks play badly."

Others charge that UN peacekeepers haven't done enough to stop rape and sexual abuse on their watch and occasionally even perpetrate abuses. Indeed, the epidemic of rape in conflict zones is shocking and horrific. That's why the United States has consistently led Security Council efforts to strengthen the mandates and means to protect civilians. That's why we pushed to create a high-level office to combat sexual violence against women and girls in conflict zones. And that's why we consistently champion accountability for genocide and justice for war criminals, whoever they are. But let's not forget the practical limitations on what peacekeepers can do. The Democratic Republic of Congo is a country the size of the United States east of the Mississippi River, with few roads, few cops, and far too many marauders. Some 20,000 peacekeepers with only a couple dozen helicopters can hardly be everywhere they may be needed all the time. Even as we demand that the UN do more and do better, we must focus our attention on the main problem: thugs with guns who deliberately use rape as a weapon of war.

Many others lament that the UN is too focused on singling out Israel. On that, they're right. UN members devote disproportionate attention to Israel and consistently adopt biased resolutions, which too often divert attention from the world's most egregious human rights abuses. I spend a good deal of time working to ensure that Israel's legitimacy is beyond dispute and its security is never in doubt. The tough issues between Israelis and Palestinians can only be solved by direct negotiations between the two parties, not in New York. We've been blunt about the deep flaws of the Goldstone Report and the Human Rights Council's inquiry into the tragic flotilla episode. We'll keep fighting to ensure that Israel has the same rights and responsibilities as all states—including membership in all appropriate regional groupings at the UN. Efforts to chip away at Israel's legitimacy will continue to be met by the frontal opposition of the United States.

Some people have criticized the Obama Administration for having sought and won a seat on the UN Human Rights Council in 2009. We have no illusions about the Human Rights Council, and we get why some people think of it as a symbol for what ails the UN. But, let me tell you this: the results were worse when America sat on the sidelines. Dictators frequently weren't called to account; abused citizens couldn't count on their voices being heard; and Israel was still bashed. We've got a long way to go to transform the Human Rights Council, but we've already gotten important results by working for real change from within. We helped set up the first-ever Special Rapporteur to monitor crackdowns on civil society groups and protect the right to free assembly and association. We twice renewed the term of the UN's Independent Expert on Sudan—the only international mechanism tracking human rights violations throughout the country. We shone the spotlight on abuses in Kyrgyzstan, Guinea, and the Democratic Republic of Congo. And, instead of abandoning Israel, we've been there to contest moves to single it out unfairly.

Put simply, some of the criticisms of the UN are overdone, and some are right on the money. Despite the UN's flaws, it's indispensable to our security in this age of tighter bonds and tighter belts. Let me provide a bit of perspective. Out of every tax dollar you pay, 34 cents goes to Social Security and Medicare, 22 cents to national security and our amazing military, and a nickel to paying interest on the national debt. Just one-tenth of a single penny goes to pay our UN dues. And here's what that buys you.

First, the UN helps prevent conflict and keep the peace around the globe. Since 1948, UN missions have saved lives, averted wars, and helped bring democracy to dozens of countries. More than 120,000 military, police, and civilian peacekeepers are now deployed in 14 operations around the world, from Haiti to Darfur to East Timor. Of those 120,000 peacekeepers, just 87 are Americans in uniform. Every peacekeeping mission must be approved by the Security Council—where America has a final say over all decisions. In Iraq and Afghanistan, UN civilian missions are mediating local disputes, coordinating international aid, and helping advance democracy—all of which helps us bring our soldiers home responsibly. UN "peacebuilding" efforts help rebuild shattered societies and prevent yesterday's hatreds from sparking tomorrow's infernos. And UN mediation has helped broker the end of conflicts, from Cambodia to Guatemala. Each UN peacekeeper costs a fraction of what it would cost to field a U.S. soldier to do the same job. So what's better, for America to bear the entire burden, or to share the burden for UN peacekeepers and pay a little more than a quarter of the cost? I don't know about you, but personally, I like places where I get 75 percent off. This is burden-sharing for a reasonable price—a lifesaving way to enable others to join us in preventing the conflict and chaos that can breed terrorism, pandemics, and other 21st-century threats. It's a whole lot more responsible to work together and share responsibility than to let threats multiply and innocents suffer.

Second, the UN helps halt the proliferation of nuclear weapons. In 2009, President Obama presided over a historic Security Council summit that unanimously adopted robust, binding steps to reduce nuclear dangers. The International Atomic Energy Agency, a key UN agency, has exposed Iran and North Korea's nuclear violations. And in the past two years, with U.S. leadership, the Security Council imposed the toughest sanctions that Iran and North Korea have ever faced. Strong Security Council resolutions have provided a foundation for others—from the European Union to Canada to South Korea—to levy additional sanctions of their own. And they warn governments that would defy their international obligations that they too will face isolation and consequences.

Third, UN humanitarian agencies go where nobody else will go to provide desperately needed food, shelter, and medicine. When polio erupted in Central Asia last year, health ministries were caught off-guard—but the World Health Organization vaccinated 6 million kids in Tajikistan and Uzbekistan, at a cost of less than $2 million. Where young people are at risk from deadly disease, UNICEF [United Nations Children's Fund] provides vaccines to fully 40 percent of the world's children and supplies millions of insecticide-treated mosquito nets in 48 countries to prevent malaria. When 125,000 Iraqi refugees were huddled in the winter chill, the UN High Commissioner for Refugees provided cash grants to buy heating fuel and warm clothes. When floods devastated Pakistan last year, the World Food Program helped feed 6.9 million people. UN humanitarian assistance doesn't just save lives. It also helps break the devastating downward spiral of chronic desperation that fuels violence and threatens international peace and security.

Fourth, the UN helps countries combat poverty, including by championing the lifesaving Millennium Development Goals. These goals include cutting extreme poverty in half and slashing the mortality rate of children under 5 by two-thirds by the year 2015. We're all more secure when people around the world have a shot at the better future we insist on for our own kids. It should trouble us deeply that half of humanity lives on less than $2.50 a day. Desperate poverty and the lack of basic services can fuel war and turmoil, creating ready havens for terrorists, criminals, and drug traffickers. Fortunately, UN development efforts afford millions the opportunity for a more dignified future. By investing in our common humanity, we simultaneously strengthen our common security.

Fifth, the UN helps foster democracy by providing expertise and oversight to strengthen fragile state institutions and support elections worldwide. Through the UN, when the people of South Sudan vote for their own freedom, the world can lend a vital hand. And when a strongman like Laurent Gbagbo of the Ivory Coast tries to steal an election, the UN, on behalf of the world, can blow the whistle.

Finally [sixth], the UN is a place where countries can come together to advance universal human rights and condemn the world's worst indignities. U.S. leadership has helped produce important results in the UN General Assembly, where we have condemned Iran, Burma, and North Korea's human rights abuses by unprecedented vote margins. We have fought and won protection for gay rights, and created UN Women, a new agency dedicated to advancing women's rights. Those steps and many more help rally the world to support bedrock U.S. and universal values: liberty, equality, and human dignity.

In the 21st century, we need the UN more than ever—to help bridge the gaps between war and reconciliation, between division and cooperation, and between misery and hope. Those of us—Democrats, Republicans, and independents alike—who support the United Nations owe it to American taxpayers to ensure that their dollars are well and cleanly spent. But, equally, those who push to curtail U.S. support to the United Nations owe it to U.S. soldiers to explain why they should perform missions now handled by United Nations peacekeepers, and they owe it to parents around the world to explain why their children should suffer without the medicine, food, and shelter that only the United Nations provides.

The United Nations plays an indispensable role in advancing our interests and defending our values. It provides a real return to the American taxpayer on our investment. The United Nations isn't perfect—far from it. The United Nations isn't the sum of our strategy—not even close. But it's an essential piece of it.

As my friend and mentor, former Secretary of State Madeleine Albright, has said, "There is a vast, sensible middle ground between those who see the United Nations as the only hope for the world and those who see in it the end of the world." A wise and deep bipartisan tradition has long seen the United Nations as essential to spurring the common actions that make Americans safer. That tradition recognizes that, if the United Nations didn't exist, we would have to invent it. Thankfully, we did help invent it. Our challenge today is to strengthen it—and in doing so, to make America more secure.

SUSAN E. RICE at the time of her speech that makes up this article was the U.S. permanent representative to the United Nations. Subsequently, on July 13, 2013, she became President Obama's national security adviser. During the presidency of Bill Clinton she served in a range of high-level posts including Assistant Secretary of State for African Affairs (1997–2001). Rice also has been a Rhodes Scholar at Oxford University, and earned a doctorate there.

Bruce S. Thornton    **NO**

The U.N.: So Bad It's Almost Beautiful

A bill introduced in Congress would allow the United States—which pays 22 percent of the United Nations' core budget and 25 percent of its peacekeeping expenses—to keep better track of how the money is spent and make sure that spending serves policies and programs consistent with American interests and principles. Yet tinkering with the United Nations' funding mechanisms will never correct the fatal flaw with the organization itself. To think otherwise is to assume that glasnost and perestroika could have saved the Soviet Union.

That flaw is the lack of consistent, unifying moral and political principles shared by member nations that can justify U.N. policies or legitimize the use of force to deter and punish aggression. Because of that absence, authoritarian, totalitarian, and even gangster regimes have seats in the U.N. General Assembly and its various councils and commissions. Of course, lip service is paid to Western ideals like universal human rights, political freedom, and liberal democracy, but these are nominally recognized not because all other nations believe in them but because of the West's economic and military dominance.

As a result, these ideals are simply redefined beyond recognition by non-Western cultures. In the Cairo Declaration on Human Rights in Islam [1990], for example, pleasing lists of "human rights" are in effect canceled out by Article 24, which says, "All the rights and freedoms stipulated in this Declaration are subject to the Islamic sharia." Or, taking their cue from Western cultural relativism, other nations dismiss such ideals as specific to the West. They argue that trying to impose Western ideals on non-Western cultures is stealth imperialism, if not outright racism.

The vacuum created by a lack of unified principles has been filled by national, political, and ideological self-interests. Thus the United Nations becomes the vehicle for pursuing those interests, as when the Soviet bloc in 1986 engineered a resolution that in effect forbade using human rights abuses as a rationale for U.N. intervention. In 1993 a U.N. conference on human rights wrote a declaration

that left out any reference to individual rights such as freedom of speech. As Israeli statesman and diplomat Dore Gold writes in *Tower of Babble*[: *How the United Nations Has Fueled Global Chaos*], "The new U.N. majority had emptied the term 'human rights' of its original meaning and hijacked it to serve its authoritarian political agenda."

Rewarding Terrorism

As the sorry history of the United Nations has shown, the various non-democratic regimes use the organization to pursue their interests at the expense of those of the United States. But then, so do American allies, as when France and Germany labored mightily in 2002 to thwart a U.N. resolution authorizing the war against Saddam Hussein, despite the fact that he had flouted seventeen previous U.N. resolutions.

The fundamental problem: member nations don't share consistent, unifying moral and political principles. Examples of such unprincipled behavior in pursuit of national interests are legion. The most egregious are the various resolutions that legitimize and reward terrorism. For example, [Palestinian leader] Yasser Arafat addressed the General Assembly wearing a holster on his hip in November 1974, a mere six months after his terrorist Palestine Liberation Organization had murdered scores of Israeli schoolchildren and three American diplomats. Arafat's visit was inevitable after the United Nations in 1970 passed Resolution 2708, which states that the United Nations "reaffirms its recognition of the legitimacy of the struggle of the colonial peoples and peoples under alien domination to exercise self-determination and independence by all the necessary means at their disposal." This free pass for terrorists was reaffirmed in 1982 when the U.N. General Assembly approved the "legitimacy of the struggle of peoples . . . from colonial and foreign domination and foreign occupation by all available means, including armed struggle."

In April 2005, the Commission on Human Rights refused to condemn killing in the name of religion. Even more despicable, in 1975—on the thirty-seventh anniversary of Kristallnacht, the Nazi pogrom against German Jews—the United Nations passed Resolution 3379, which defined Zionism as a form of racism. This odious resolution was revoked sixteen years later, but only because Israel had made its repeal a condition of participating in the Madrid peace conference. That this repeal reflected expediency rather than principle was obvious in April 2002, when the U.N. Commission on Human Rights affirmed "the legitimate right of the Palestinian people to resist Israeli occupation," just after a Hamas suicide bomber had killed thirty Israelis celebrating Passover.

The Commission on Human Rights and its allegedly improved successor, the Human Rights Council, may be the best representatives of the United Nations' Orwellian hypocrisy [a reference to George Owell's *1984*, a novel in which the government Ministry of Truth changes historical records to reflect the doctrine of the government]. Thug states like Iran, Sudan, Cuba, China, Zimbabwe, Saudi Arabia, and North Korea, which support terrorism and violate human rights as a matter of policy, have been allowed to sit on the council, where terrorist and state violence is never censured, even as Israel faces serial condemnation. Indeed, in April 2005, the commission refused to condemn killing in the name of religion. At the same time, it asserted that criticizing Muslim terrorists was "defamation of religion."

In March 2007, the council's response to the killings and riots that followed the publication of cartoons depicting Muhammad was to call for a ban on the defamation of religion—even as it ignored the threat to the human right to free speech. Neither the genocidal charter of Hamas [a Palestinian political organization considered to be terrorist by the United States and many other countreis] nor the widespread, state-sanctioned anti-Semitism in the Middle East has ever been condemned, while thirty-three resolutions through 2010 have criticized Israel. The animus of the Human Rights Council against the only liberal democracy in the Middle East was evident recently in its wildly inaccurate and biased Goldstone Report, an investigation into Israel's actions in Gaza. [The Goldstone Report is a 575-page document, *Report of the United Nations Fact Finding Mission on the Gaza Conflict* submitted in 2009 to the Human Rights Council by a UN investigative commission headed by South African Judge Richard Goldstone.] Even the report's author was compelled to disavow it because of inaccuracies and obvious bias. Like the United Nations, the council is an instrument of member states' interests, not the presumed principles and rights enshrined in its charter and rhetoric.

"Harmony of Interests" a Cruel Illusion

Given its purpose as a means for weak or autocratic states to pursue their interests, the United Nations has evolved into a bloated, corrupt, ineffective bureaucracy. Its budget has doubled since 2000. The most famous U.N. scandal is the 1995–2003 oil-for-food program that operated in Saddam Hussein's Iraq, overseeing $15 billion a year supposedly meant to feed the Iraqi people. Instead it was "an open bazaar of payoffs, favoritism, and kickbacks," as *The New York Times* put it, generating over $10 billion in illicit funds for Saddam's regime and billions more for Russian and French politicians and businessmen.

Worse than these financial scandals, however, is the utter impotence of the United Nations in stopping violence in places like Sudan, Bosnia, and Rwanda, where horrific violence occurred a stone's throw away from U.N. "peacekeeping" forces. In fact, in Bosnia, U.N. "safe areas" simply made it easier for the Serbs to round up and slaughter seven thousand Bosnian Muslims.

Even worse than its financial scandals is the U.N.'s utter impotence in stopping violence in places like Sudan, Bosnia, and Rwanda. The United Nations is a relic of the same Enlightenment idealism that has driven internationalism for almost two centuries and which has failed dismally to stop the violence of the twentieth century and beyond. That idealism assumes that all humanity is progressing beyond the use of force, tyrannical regimes, and parochial nationalist interests to a transnational "harmony of interests" created by communication technologies, global trade, and the spread of liberal democracy. These shared interests, moreover, can be institutionalized in international laws, courts, treaties, and supranational organizations that will substitute diplomacy and negotiation for force.

This vision created the League of Nations, which in the Twenties and Thirties completely failed to stop the state violence of Japan, Italy, and Germany. Its successor, the United Nations, has done no better for the simple reason that such a "harmony of interests" does not exist and never will. States and peoples have different values, beliefs, and aims. They pursue interests that conflict with the interests and aims of other states. And the melancholy lesson of history is that these conflicts usually are resolved by force or the credible threat of force, not by diplomatic chatter in a "cockpit in the Tower of Babel," to use the phrase conjured up by Winston Churchill when the United Nations was born.

The question, then, is not how we fix the United Nations, as the U.N. Transparency, Accountability, and

Reform Act, introduced in August [2011] by Florida Republican Ileana Ros-Lehtinen [chairwoman of the Committee on Foreign Affairs, U.S. House of Representatives], attempts to do. Instead it is why we continue to spend U.S. taxpayer dollars—$7.7 billion in 2010—on an institution filled with states hostile to us and working against our own foreign policy interests. Herein lies the greatest flaw in the thinking of those Americans who still believe in the usefulness of the United Nations: they believe that unelected, unaccountable functionaries of tyrannous regimes—regimes not only pursuing their own interests but frequently working against our interests—are more capable of determining the legitimacy of the United States' foreign policy and behavior than are the American people.

In contrast to the United Nations, the legitimacy of American actions is conferred by the democratic process:

the free, open debate on the part of citizens who can hold their leaders accountable and have a sense of the ideals and principles that animate foreign policy and provide its goals. Subjecting those decisions to the corrupt deliberations of the United Nations merely hampers our own interests and endangers our national security. We need to get out of the United Nations, not fix it.

Bruce S. Thornton is a research fellow at the Hoover Institution and a professor of classics and humanities at California State University in Fresno, California. He received a PhD degree in comparative literature from the University of California at Los Angeles. His most recent book is *The Wages of Appeasement: Ancient Athens, Munich, and Obama's America* (Encounter Books, 2011).

EXPLORING THE ISSUE

Is the UN a Worthwhile Organization?

Critical Thinking and Reflection

1. Both Susan Rice and Bruce Thornton in the YES and NO selections evaluate the UN in terms of how well it advances U.S. policy and interests. Is that a correct standard, even for Americans, or should the UN be evaluated from the point of view of how well it promotes the goals of its Charter including maintaining and promoting peace, protecting human rights, and improving the conditions of less fortunate people and countries?
2. To fund its operations, the UN assesses countries based on their wealth. The United States pays the greatest share, which came to about $3 billion or general operations and peacekeeping in 2012. The U.S. also contributes about another $5 billion to various UN-associated agencies like the World Health Organization. These funds total about two-tenths of 1 percent of the U.S. budget. Some say these amounts to the UN are too high. Do you agree?
3. Would you favor or oppose abolishing the veto in the UN Security Council?

Is There Common Ground?

Evaluating the value of the UN has a great deal to do with what standard of evaluation you adopt. One standard is akin to asking whether a glass is half full or half empty. Undoubtedly, the UN has not come anywhere near achieving the lofty goals set out in its Charter. Thus, the evaluative glass is at least half empty. However, it is also true that the UN has accomplished a great deal. Many UN peacekeeping operations have been fielded, and some of them have made an important contribution. In late 2012, there were over 97,000 troops, police, and other personnel from 116 countries trying to provide security through 18 different UN peacekeeping operations. For all the human abuse and poverty that remain, there are greater justice and better living conditions in the world now than that existed a few decades ago. Again, the UN can legitimately claim some of the credit. So from this perspective, the glass is at least fuller than it once was.

Yet another standard of measurement has to do with the old adage about being careful when throwing stones in glass houses. The UN has sometimes wasted money, its workers have not always performed admirably, and even Secretary General Kofi Annan (1997–2006) conceded that to some degree the UN's administration had "become fragmented, duplicative, and rigid." Yet every government, including the U.S. government, is subject to the same accusation. Annan was able to implement many changes, and his successor beginning in 2006, Secretary General

Ban Ki-moon, has said continued reform will be a top priority. It is also the case that a small number of UN troops and personnel have behaved immorally, but the equally if not more abominable performance of a few American soldiers at Abu Ghraib Prison and elsewhere show how the actions of the reprehensive few can besmirch the reputations of the honorable many.

The ultimate question is if not the UN, then what. Calls to "fix" the UN are many and varied, and debating what should be done and what is possible are worthwhile. Perhaps creating a successor organization would be a good idea. Then there is the suggestion of doing without a central global organization. Ask yourself whether that would open the way to a safer, more prosperous, more just world or even further your country's national interests.

Create Central

www.mhhe.com/createcentral

Additional Resources

Mingst, Karen A., and Karns, Margaret P. (2011). *The United Nations in the 21st Century* (Westview Press)

United Nations. (2014). *Basic Facts about the United Nations* (United Nations)

Weiss, Thomas G. (2013). *What's Wrong with the United Nations and How to Fix It* (John Wiley & Sons)

Internet References . . .

United Nations Foundation

http://www.unfoundation.org/

United States Mission to the United Nations

www.usunnewyork.usmission.gov/

Welcome to the United Nations

www.un.org/en

Selected, Edited, and with Issue Framing Material by:
John T. Rourke, *University of Connecticut, Storrs*

ISSUE

Is U.S. Refusal to Join the International Criminal Court Justifiable?

YES: Brett Schaefer and Steven Groves, from "The U.S. Should Not Join the International Criminal Court," Backgrounder on International Organization, The Heritage Foundation (2009)

NO: Michael P. Scharf, from "Is a U.N. International Criminal Court in the U.S. National Interest?" Testimony Before the Subcommittee on International Operations of the Committee on Foreign Relations, U.S. Senate (1998)

Learning Outcomes

After reading this issue, you will be able to:

- Explain the background of the international court of justice (ICJ).
- Indicate what the U.S. objections are to the adhering to the Rome Treaty and joining the ICJ.
- Enumerate the counterarguments for joining the ICJ.

ISSUE SUMMARY

YES: Brett Schaefer, the Jay Kingham fellow in international regulatory affairs at the Heritage Foundation, and Steven Groves, the Bernard and Barbara Lomas fellow at the Margaret Thatcher Center for Freedom, a division of the Kathryn and Shelby Cullom Davis Institute for International Studies at the Heritage Foundation, contend that although the court's supporters have a noble purpose, there are a number of reasons to be cautious and concerned about how ratification of the Rome Statute would affect U.S. sovereignty and how ICC action could affect politically precarious situations around the world.

NO: Michael P. Scharf, a professor of law at and director of the Center for International Law and Policy, New England School of Law, argues in testimony given just after the establishment of the ICC was finalized by the Conference at Rome and sent to the world's countries for adoption (ratification) that while the United States did not get everything it wanted in the Treaty of Rome creating the ICC, it is a worthwhile step forward toward global justice.

Historically, international law has focused primarily on countries. More recently, individuals have increasingly become subject to international law. The first major step in this direction was the convening of the Nuremberg and Tokyo war crimes trials after World War II to try German and Japanese military and civilian leaders charged with various war crimes. There were no subsequent war crimes tribunals until the 1990s when the United Nations (UN) established two of them. One sits in The Hague, the Netherlands, and deals with the horrific events in Bosnia. The other tribunal is in Arusha, Tanzania, and provides justice for the genocidal massacres in Rwanda. These tribunals have indicted numerous people for war crimes and have convicted and imprisoned many of them. Nevertheless, there was a widespread feeling that such ad hoc tribunals needed to be replaced by a permanent international criminal tribunal.

In 1996, the UN convened a conference in Rome to do just that. At first the United States was supportive, but it favored a very limited court that could only prosecute and hear cases referred to it by the UN Security Council (where the United States had a veto) and, even then, could only try individuals with the permission of the defendant's home government. Most countries disagreed, but in 1998 the Rome conference voted overwhelmingly to create a relatively strong court. The Rome Statute of the International Criminal Court (ICC) gives the ICC jurisdiction over wars of aggression, genocide, and other crimes, but only if the home country of an alleged perpetrator fails to act.

Although the ICC treaty was open for signature in July 1998, President Bill Clinton showed either ambivalence or a desire not to have it injected as an issue into the 2000 presidential election by waiting until December 31, 2000, to have a U.S. official sign the treaty. If Clinton had his doubts, his successor, George W. Bush, did not. He was adamantly opposed to the treaty. As directed by the White House, State Department official John R. Bolton sent a letter dated May 6, 2002, to UN Secretary General Kofi Annan informing him that "in connection with the Rome Statute of the International Criminal Court . . . , the United States does not intend to become a party to the treaty . . . [and] has no legal obligations arising from its signature on December 31, 2000." The Bush administration also launched an effort to persuade other countries to sign "Article 98" agreements by which countries agree not to surrender U.S. citizens to the ICC.

The letter formally notifying the UN that the United States does not intend to become a party to the Rome Statute also ended any U.S. participation in the workings of the court. In the YES selection Brett Schaefer and Steven Groves review the ICC's legal implications and assess its operation so far. As a result, they conclude that although the court reflects an admirable desire to hold war criminals accountable for their terrible crimes, the ICC is so flawed that it would be unwise for the United States to join it. In the NO selection Michael Scharf takes the view that while the ICC may cause some concerns for Americans, it is a major advance in world justice that puts potential war criminals on notice that starting wars and violating the rules governing the treatment of civilians as well as combatants may lead to being charged and tried by an impartial international court and, if convicted, sentenced to prison." While Sharf's testimony was given soon after the founding of the ICC treaty, his comments are as relevant now as they were then.

YES

Brett Schaefer and Steven Groves

The U.S. Should Not Join the International Criminal Court

The idea of establishing an international court to prosecute serious international crimes—war crimes, crimes against humanity, and genocide—has long held a special place in the hearts of human rights activists and those hoping to hold perpetrators of terrible crimes to account. In 1998, that idea became reality when the Rome Statute of the International Criminal Court was adopted at a diplomatic conference convened by the U.N. General Assembly. The International Criminal Court (ICC) was formally established in 2002 after 60 countries ratified the statute. The ICC was created to prosecute war crimes, crimes against humanity, genocide, and the as yet undefined crime of aggression. Regrettably, although the court's supporters have a noble purpose, there are a number of reasons to be cautious and concerned about how ratification of the Rome Statute would affect U.S. sovereignty and how ICC action could affect politically precarious situations around the world.

Among other concerns, past U.S. Administrations concluded that the Rome Statute created a seriously flawed institution that lacks prudent safeguards against political manipulation, possesses sweeping authority without accountability to the U.N. Security Council, and violates national sovereignty by claiming jurisdiction over the nationals and military personnel of non-party states in some circumstances. These concerns led President Bill Clinton to urge President George W. Bush not to submit the treaty to the Senate for advice and consent necessary for ratification. After extensive efforts to change the statute to address key U.S. concerns failed, President Bush felt it necessary to "un-sign" the Rome Statute by formally notifying the U.N. Secretary-General that the U.S. did not intend to ratify the treaty and was no longer bound under international law to avoid actions that would run counter to the intent and purpose of the treaty. Subsequently, the U.S. took a number of steps to protect its military personnel, officials, and nationals from ICC claims of jurisdiction.

Until these and other concerns are fully addressed, the Obama Administration should resist pressure to "re-sign" the Rome Statute, eschew cooperation with the ICC except when

U.S. interests are affected, and maintain the existing policy of protecting U.S. military personnel, officials, and nationals from the court's illegitimate claims of jurisdiction. Nor should the Obama Administration seek ratification of the Rome Statute prior to the 2010 review, and then only if the Rome Statute and the ICC and its procedures are amended to address all of the serious concerns that led past U.S. Administrations to oppose ratification of the Rome Statute.

Background

The United States has long championed human rights and supported the ideal that those who commit serious human rights violations should be held accountable. Indeed, it was the United States that insisted—over Soviet objections—that promoting basic human rights and fundamental freedoms be included among the purposes of the United Nations. The United States also played a lead role in championing major international efforts in international humanitarian law, such as the Geneva Conventions.

The U.S. has supported the creation of international courts to prosecute gross human rights abuses. It pioneered the Nuremburg and Tokyo tribunals to prosecute atrocities committed during World War II. Since then, the U.S. was a key supporter of establishing the ad hoc International Criminal Tribunal for the former Yugoslavia (ICTY) and International Criminal Tribunal for Rwanda (ICTR), which were both approved by the Security Council.

Continuing its long support for these efforts, the U.S. initially was an eager participant in the effort to create an International Criminal Court in the 1990s. However, once negotiations began [in 1998] on the final version of the Rome Statute, America's support waned because many of its concerns were ignored or opposed outright. According to David J. Scheffer, chief U.S. negotiator at the 1998 Rome conference:

> In Rome, we indicated our willingness to be flexible. . . . Unfortunately, a small group of countries, meeting behind closed doors in the final days of

the Rome conference, produced a seriously flawed take-it-or-leave-it text, one that provides a recipe for politicization of the court and risks deterring responsible international action to promote peace and security.

In the end, despite persistent efforts to amend the Rome Statute to alleviate U.S. concerns, the conference rejected most of the changes proposed by the U.S., and the final document was approved over U.S. opposition.

Since the approval of the Rome Statute in 1998, U.S. policy toward the ICC has been clear and consistent: The U.S. has refused to join the ICC because it lacks prudent safeguards against political manipulation, possesses sweeping authority without accountability to the U.N. Security Council, and violates national sovereignty by claiming jurisdiction over the nationals and military personnel of non-party states in some circumstances.

The United States is not alone in its concerns about the ICC. As of August 6, 2009, only 110 of the 192 U.N. member states had ratified the Rome Statute. In fact, China, India, and Russia are among the other major powers that have refused to ratify the Rome Statute out of concern that it unduly infringes on their foreign and security policy decisions—issues rightly reserved to sovereign governments and over which the ICC should not claim authority.

The ICC's Record

The International Criminal Court has a clear legal lineage extending back to the Nuremburg and Tokyo trials and ad hoc tribunals, such as the ICTY and the ICTR, which were established by the U.N. Security Council in 1993 and 1994, respectively. However, the ICC is much broader and more independent than these limited precedents. Its authority is not limited to disputes between governments as is the case with the International Court of Justice (ICJ) or to a particular jurisdiction as is the case with national judiciaries. Nor is its authority limited to particular crimes committed in a certain place or period of time as was the case with the post-World War II trials and the Yugoslavian and Rwandan tribunals.

Instead, the ICC claims jurisdiction over individuals committing genocide, crimes against humanity, war crimes, and the undefined crime of aggression. This jurisdiction extends from the entry into force of the Rome Statute in July 2002 and applies to all citizens of states that have ratified the Rome Statute. However, it also extends to individuals from countries that are not party to the Rome Statute if the alleged crimes occur on the territory of an ICC party state, the non-party government invites ICC jurisdiction, or the U.N. Security Council refers the case to the ICC.

International lawyers Lee Casey and David Rivkin point out that the ICC is a radical departure from previous international courts [because] "It has jurisdiction over individuals, including elected or appointed government officials, and its judgments may be directly enforced against them, regardless of their own national constitutions or court systems."

> Moreover, the court's structure establishes few, if any, practical external checks on the ICC's authority. Among the judges' responsibilities are determining whether the prosecutor may proceed with a case and whether a member state has been "unwilling or unable genuinely to carry out the investigation or prosecution," which would trigger the ICC's jurisdiction under the principle of "complementarity," which is designed to limit the court's power and avoid political abuse of its authority. Thus, the various arms of the ICC are themselves the only real check on its authority.

Even though the Rome Statute entered into force in July 2002, there is little concrete basis for judging the ICC's performance. Shortly after its formal establishment, the ICC began receiving its first referrals. Currently, the ICC has opened four cases, involving situations in the Democratic Republic of Congo (DRC), Uganda, the Central African Republic, and Darfur, Sudan.

As an institution, the ICC has performed little, if any, better than the ad hoc tribunals that it was created to replace. Like the Rwandan and Yugoslavian tribunals, the ICC is slow to act. The ICC prosecutor took six months to open an investigation in Uganda, two months with the DRC, over a year with Darfur, and nearly two years with the Central African Republic. It has yet to conclude a full trial cycle more than seven years after being created. Moreover, like the ad hoc tribunals, the ICC can investigate and prosecute crimes only after the fact. The alleged deterrent effect of a standing international criminal court has not ended atrocities in the DRC, Uganda, the Central African Republic, or Darfur, where cases are ongoing. Nor has it deterred atrocities by Burma against its own people, crimes committed during Russia's 2008 invasion of Georgia (an ICC party), ICC party Venezuela's support of leftist guerilllas in Colombia, or any of a number of other situations around the world where war crimes or crimes against humanity may be occurring.

Another problem is that the ICC lacks a mechanism to enforce its rulings and is, therefore, entirely dependent on governments to arrest and transfer perpetrators to the court. However, such arrests can have significant diplomatic consequences, which can greatly inhibit the efficacy of the court in pursuing its warrants and prosecuting outstanding cases. The most prominent example is Sudanese President Bashir's willingness to travel to other countries on official visits—thus far only to non-ICC states—despite the ICC arrest warrant. This flaw was also present with the ICTY and the ICTR, although they could at least rely on a Security Council resolution mandating international cooperation in enforcing their arrest warrants. In contrast, the Nuremburg and Tokyo tribunals were established where the authority of the judicial proceedings could rely on Allied occupation forces to search out, arrest, and detain the accused.

The Myth of Bush Administration Intransigence

The U.S. refusal to ratify the Rome Statute has been mischaracterized by ICC proponents as solely a Bush Administration policy. In fact, the Clinton Administration initiated the U.S. policy of distancing itself from the ICC. According to David J. Scheffer, Ambassador-at-Large for War Crimes Issues under the Clinton Administration:

> Foreign officials and representatives of non-governmental organizations tried to assure us in Rome that procedural safeguards built into the treaty—many sought successfully by the United States—meant that there would be no plausible risk to U.S. soldiers. We could not share in such an optimistic view of the infallibility of an untried institution. . . .

President Clinton himself acknowledged the treaty's "significant flaws" and recommended that President Bush not submit the treaty to the Senate for advice and consent. When President Clinton authorized the U.S. delegation to sign the Rome Statute on December 31, 2000, it was not to pave the way for U.S. ratification, but solely to give the U.S. an opportunity to address American concerns about the ICC. As Clinton said at the time in his signing statement:

> In signing, however, we are not abandoning our concerns about significant flaws in the treaty. In particular, we are concerned that when the court comes into existence, it will not only exercise authority over personnel of states that have ratified the treaty but also claim jurisdiction over personnel of states that have not. With signature, however, we will be in a position to influence the evolution of the court. Without signature, we will not.

After adoption of the Rome Statute in 1998, both the Clinton and Bush Administrations sought to rectify the parts of the statute that precluded U.S. participation. Specifically,

the U.S. actively participated in the post-Rome preparatory commissions, hoping to address its concerns. As former U.S. Under Secretary for Political Affairs Marc Grossman noted:

> After the United States voted against the treaty in Rome, the U.S. remained committed and engaged—working for two years to help shape the court and to seek the necessary safeguards to prevent a politicization of the process. While we were able to make some improvements during our active participation in the UN Preparatory Commission meetings in New York, we were ultimately unable [to] obtain the remedies necessary to overcome our fundamental concerns. . . .

The consequences of failing to change the objectionable provisions of the Rome Statute became acute when the 60th country ratified the treaty, causing the statute to enter into force in July 2002. Faced with the prospect of a functioning International Criminal Court that could assert jurisdiction over U.S. soldiers and officials in certain circumstances, the Bush Administration and Congress took steps to protect Americans from the court's jurisdiction, which the U.S. did not recognize. For instance, Congress passed the American Service-Members' Protection Act of 2002 (ASPA), which restricts U.S. interaction with the ICC and its state parties by:

- Prohibiting cooperation with the ICC by any official U.S. entity, including providing support or funds to the ICC, extraditing or transferring U.S. citizens or permanent resident aliens to the ICC, or permitting ICC investigations on U.S. territory.
- Prohibiting participation by U.S. military or officials in U.N. peacekeeping operations unless they are shielded from the ICC's jurisdiction.
- Prohibiting the sharing of classified national security information or other law enforcement information with the ICC.
- Constraining military assistance to ICC member states, except NATO countries and major non-NATO allies and Taiwan, unless they entered into an agreement with the U.S. not to surrender U.S. persons to the ICC without U.S. permission.
- Authorizing the President to use "all means necessary and appropriate" to free U.S. military personnel or officials detained by the ICC.

Congress also approved the Nethercutt Amendment to the foreign operations appropriations bill for fiscal year 2005, which prohibited disbursement of selected U.S. assistance to an ICC party unless the country has entered into a bilateral agreement not to surrender U.S. persons to the ICC (commonly known as an Article 98 agreement) or is specifically exempted in the legislation. Both ASPA and the Nethercutt Amendment contained waiver provisions allowing the President to ignore these restrictions with notification to Congress. In recent years, Congress has repealed or loosened restrictions on providing assistance to ICC state parties that have not entered into Article 98 agreements with the U.S. However, other ASPA restrictions remain in effect.

The Bush Administration signed these legislative measures and undertook several specific efforts to fulfill the mandates of the legislation and to protect U.S. military personnel and officials from potential ICC prosecution.

Possible Legal Obligations from Signing the Rome Statute

Under Article 18 of the Vienna Convention on the Law of Treaties, the Bush Administration determined that its efforts to protect U.S. persons from the ICC could be construed as "acts which would defeat the object and purpose of a treaty." To resolve this potential conflict, the U.S. sent a letter to U.N. Secretary-General Kofi Annan, the depositor for the Rome Statute, stating that it did not intend to become a party to the Rome Statute and declaring that "the United States has no legal obligations arising from its signature" of the Rome Statute. This act has been described as "un-signing" the Rome Statute. As John Bellinger, former Legal Advisor to Secretary of State Condoleezza Rice, made clear in a 2008 speech, "the central motivation was to resolve any confusion whether, as a matter of treaty law, the United States had residual legal obligations arising from its signature of the Rome Statute."

Article 98 Agreements

Because the ICC could claim jurisdiction over non-parties to the Rome Statute—an assertion unprecedented in international legal jurisdiction—the Bush Administration sought legal protections to preclude nations from surrendering, extraditing, or transferring U.S. persons to the ICC or third countries for that purpose without U.S. consent. Under an Article 98 agreement, a country agrees not to turn U.S. persons over to the ICC without U.S. consent.

Contrary to the claims of the more strident critics, who label the Article 98 agreements as "bilateral immunity agreements" or "impunity agreements," the agreements neither absolve the U.S. of its obligation to investigate and prosecute alleged crimes, constrain the other nation's ability to investigate and prosecute crimes committed by an American person within its jurisdiction, nor constrain an international tribunal established by the Security Council

from investigating or prosecuting crimes committed by U.S. persons. The agreements simply prevent other countries from turning U.S. persons over to an international court that does not have jurisdiction recognized by the United States.

The limited nature of the agreements is entirely consistent with international law, which supports the principle that a state cannot be bound by a treaty to which it is not a party. The agreements are also consistent with customary international law because the issue of ICC jurisdiction is very much in dispute. Moreover, they are consistent with the Rome Statute itself, which contemplates such agreements in Article 98:

> The Court may not proceed with a request for surrender which would require the requested State to act inconsistently with its obligations under international agreements pursuant to which the consent of a sending State is required to surrender a person of that State to the Court, unless the Court can first obtain the cooperation of the sending State for the giving of consent for the surrender.

Although the U.S. is not currently seeking to negotiate additional Article 98 agreements, there are no known plans to terminate existing agreements. Reportedly, 104 countries have signed Article 98 agreements with the U.S., of which 97 agreements remain in effect.

Language to Protect U.S. Persons

In 2002, the U.S. sought a Security Council resolution to indefinitely exempt from ICC jurisdiction U.S. troops and officials participating in U.N. peacekeeping operations. The effort failed in the face of arguments that the Security Council lacked the authority to rewrite the terms of the Rome Statute, but the Security Council did adopt Resolution 1422, which deferred ICC prosecution of U.N. peacekeeping personnel for one year under Article 16 of the Rome Statute. The deferral was renewed once and expired in June 2004. The U.S. also successfully included language in Resolution 1497 on the U.N. Mission to Liberia granting exclusive jurisdiction over "current or former officials or personnel from a contributing State" to the contributing state if it is not a party to the Rome Statute.

Persistent Barriers to U.S. Ratification

ICC supporters have called for the Obama Administration to re-sign the Rome Statute, reverse protective measures secured during the Bush Administration (Article 98 agreements), and fully embrace the ICC. Indeed, the Obama Administration may be considering some or all of those actions. However, the ICC's flaws advise caution and concern, particularly in how the ICC could affect national sovereignty and politically precarious situations around the globe.

When it decided to un-sign the Rome Statute, the Bush Administration voiced five concerns regarding the Rome Statute. These critical concerns have not been addressed.

The ICC's Unchecked Power

The U.S. system of government is based on the principle that power must be checked by other power or it will be abused and misused. With this in mind, the Founding Fathers divided the national government into three branches, giving each the means to influence and restrain excesses of the other branches. For instance, Congress confirms and can impeach federal judges and has the sole authority to authorize spending, the President nominates judges and can veto legislation, and the courts can nullify laws passed by Congress and overturn presidential actions if it judges them unconstitutional.

> The ICC lacks robust checks on its authority, despite strong efforts by U.S. delegates to insert them during the treaty negotiations. The court is an independent treaty body. In theory, the states that have ratified the Rome Statute and accepted the court's authority control the ICC. In practice, the role of the Assembly of State Parties is limited. The judges themselves settle any dispute over the court's "judicial functions." The prosecutor can initiate an investigation on his own authority, and the ICC judges determine whether the investigation may proceed. The U.N. Security Council can delay an investigation for a year—a delay that can be renewed—but it cannot stop an investigation.

The Challenges to the Security Council's Authority

The Rome Statute empowers the ICC to investigate, prosecute, and punish individuals for the as yet undefined crime of "aggression." This directly challenges the authority and prerogatives of the U.N. Security Council, which the U.N. Charter gives "primary responsibility for the maintenance of international peace and security" and which is the only U.N. institution empowered to determine when a nation has committed an act of aggression. Yet, the Rome Statute "empowers the court to decide on this matter and lets the prosecutor investigate and prosecute this undefined crime" free of any oversight from the Security Council.

A Threat to National Sovereignty

A bedrock principle of the international system is that treaties and the judgments and decisions of treaty organizations cannot be imposed on states without their consent. In certain circumstances, the ICC claims the authority to detain and try U.S. military personnel, U.S. officials, and other U.S. nationals even though the U.S. has not ratified the Rome Statute and has declared that it does not consider itself bound by its signature on the treaty. As Grossman noted, "While sovereign nations have the authority to try non-citizens who have committed crimes against their citizens or in their territory, the United States has never recognized the right of an international organization to do so absent consent or a U.N. Security Council mandate."

As such, the Rome Statute violates international law as it has been traditionally understood by empowering the ICC to prosecute and punish the nationals of countries that are not party to it. In fact, Article 34 of the Vienna Convention on the Law of Treaties unequivocally states: "A treaty does not create either obligations or rights for a third State without its consent."

> Protestations by ICC proponents that the court would seek such prosecutions only if a country is unwilling or unable to prosecute those accused of crimes within the court's jurisdiction—the principle of complementarity—are insufficient to alleviate sovereignty concerns.

For example, the Obama Administration recently declared that no employee of the Central Intelligence Agency (CIA) who engaged in the use of "enhanced interrogation techniques" on detainees would be criminally prosecuted. That decision was presumably the result of an analysis of U.S. law, legal advice provided to the CIA by Justice Department lawyers, and the particular actions of the interrogators. Yet if the U.S. were a party to the Rome Statute, the Administration's announced decision not to prosecute would fulfill a prerequisite for possible prosecution by the ICC under the principle of complementarity. That is, because the U.S. has no plans to prosecute its operatives for acts that many in the international community consider torture, the ICC prosecutor would be empowered (and possibly compelled) to pursue charges against the interrogators.

Erosion of Fundamental Elements of the U.N. Charter

The ICC's jurisdiction over war crimes, crimes against humanity, genocide, and aggression directly involves the court in fundamental issues traditionally reserved to sovereign states, such as when a state can lawfully use armed force to defend itself, its citizens, or its interests; how and to what extent armed force may be applied; and the point at which particular actions constitute serious crimes. Blurring the lines of authority and responsibility in these decisions has serious consequences. As Grossman notes, "with the ICC prosecutor and judges presuming to sit in judgment of the security decisions of States without their assent, the ICC could have a chilling effect on the willingness of States to project power in defense of their moral and security interests." The ability to project power must be protected, not only for America's own national security interests, but also for those individuals threatened by genocide and despotism who can only be protected through the use of force.

Complications to Military Cooperation Between the U.S. and Its Allies

The treaty creates an obligation to hand over U.S. nationals to the court, regardless of U.S. objections, absent a competing obligation such as that created through an Article 98 agreement. The United States has a unique role and responsibility in preserving international peace and security. At any given time, U.S. forces are located in approximately 100 nations around the world, standing ready to defend the interests of the U.S. and its allies, engaging in peacekeeping and humanitarian operations, conducting military exercises, or protecting U.S. interests through military intervention. The worldwide extension of U.S. armed forces is internationally unique. The U.S. must ensure that its soldiers and government officials are not exposed to politically motivated investigations and prosecutions.

Ongoing Causes for Concern

Supporters of U.S. ratification of the Rome Statute often dismiss these concerns as unjustified, disproved by the ICC's conduct during its first seven years in operation, or as insufficient to overcome the need for an international court to hold perpetrators of serious crimes to account. Considering the other options that exist or could be created to fill the ICC's role of holding perpetrators of war crimes, crimes against humanity, genocide, and aggression to account, the benefits from joining such a flawed institution do not justify the risks.

Furthermore, based on the ICC's record and the trend in international legal norms, they are being disingenuous in dismissing concerns about overpoliticization of the ICC, its impact on diplomatic initiatives and sovereign decisions on the use of force, its expansive claim of jurisdiction over the citizens of non-states parties, and incompatibility with U.S. legal norms and traditions. A number of specific risks are obvious.

Politicization of the Court

Unscrupulous individuals and groups and nations seeking to influence foreign policy and security decisions of other nations have and will continue to seek to misuse the ICC for politically motivated purposes. Without appropriate checks and balances to prevent its misuse, the ICC represents a dangerous temptation for those with political axes to grind. The prosecutor's *proprio motu* authority to initiate an investigation based solely on his own authority or on information provided by a government, a nongovernmental organization (NGO), or individuals is an open invitation for political manipulation.

One example is the multitude of complaints submitted to the ICC urging the court to indict Bush Administration officials for alleged crimes in Iraq and Afghanistan. The Office of the Prosecutor received more than 240 communications alleging crimes related to the situation in Iraq. Thus far, the prosecutor has demonstrated considerable restraint, declining to pursue these cases for various reasons, including that the ICC does not have "jurisdiction with respect to actions of non-State Party nationals on the territory of Iraq," which is also not a party to the Rome Statute.

All current ICC cases were referred to the ICC by the governments of the territories in which the alleged crimes occurred or by the Security Council. Comparatively speaking, these cases are low-hanging fruit—situations clearly envisioned to be within the authority of the court by all states. Even so, they have not been without controversy, as demonstrated by the AU reaction to the arrest warrant for President Bashir and attempts to have the Security Council defer the case.

However, the ICC's brief track record is no assurance that future cases will be similarly resolved, especially given the increasing appetite for lodging charges with the ICC. A far more significant test will arise if the prosecutor decides to investigate (and the court's pre-trial chamber authorizes) a case involving a non-ICC party without a Security Council referral or against the objections of the government of the involved territory.

This could arise from the prosecutor's monitoring of the situation in Palestine. Even though Israel is not a party to the Rome Statute, the ICC prosecutor is exploring a request by the Palestinian National Authority to prosecute Israeli commanders for alleged war crimes committed during the recent actions in Gaza. The request is supported by 200 complaints from individuals and NGOs alleging war crimes by the Israeli military and civilian leaders related to military actions in Gaza.

Palestinian lawyers maintain that the Palestinian National Authority can request ICC jurisdiction as the de facto sovereign even though it is not an internationally recognized state. By countenancing Palestine's claims, the ICC prosecutor has enabled pressure to be applied to Israel over alleged war crimes, while ignoring Hamas's incitement of the military action and its commission of war crimes against Israeli civilians. Furthermore, by seemingly recognizing Palestine as a sovereign entity, the prosecutor's action has arguably created a pathway for Palestinian statehood without first reaching a comprehensive peace deal with Israel. This determination is an inherently political issue beyond the ICC's authority, yet the prosecutor has yet to reject the possibility that the ICC may open a case on the situation.

Alternatively, the prosecutor could raise ire by making a legal judgment call on a crime under the court's jurisdiction that lacks a firm, universal interpretation, such as:

- "Committing outrages upon personal dignity, in particular humiliating and degrading treatment";
- "Intentionally launching an attack in the knowledge that such attack will cause incidental loss of life or injury to civilians or damage to civilian objects or widespread, long-term and severe damage to the natural environment which would be clearly excessive in relation to the concrete and direct overall military advantage anticipated"; or
- Using weapons "which are of a nature to cause superfluous injury or unnecessary suffering or which are inherently indiscriminate in violation of the international law of armed conflict."

In each of these cases, a reasonable conclusion could be made to determine whether a crime was committed. For instance, many human rights groups allege outrages on personal dignity and "humiliating and degrading treatment" were committed at the detention facility at Guantanamo Bay, Cuba. The U.S. disputes these claims. Excessive use of force has been alleged in Israel's attacks in Gaza, while others insist Israel demonstrated forbearance and consideration in trying to prevent civilian casualties. There is also an ongoing international effort to ban landmines and cluster munitions. If the ICC member states agree to add them to the annex of banned weapons, it could lead to a confrontation over their use by non-party states, such as the U.S., which opposes banning these weapons. These are merely some scenarios in which politicization could become an issue for the ICC.

The Undefined Crime of Aggression

It would be irresponsible for the U.S. to expose its military personnel and civilian officials to a court that has yet to define the very crimes over which it claims jurisdiction. Yet that is the situation the U.S. would face if it ratified the Rome Statute. The Statute includes the crime of aggression as one of its enumerated crimes, but the crime has yet to be defined, despite a special working group that has been debating the issue for more than five years.

For instance, some argue that any military action conducted without Security Council authorization violates international law and is, therefore, an act of aggression that could warrant an ICC indictment. The U.S. has been the aggressor in several recent military actions, including military invasions of the sovereign territories of Afghanistan and Iraq, albeit with the U.N. Security Council's blessing in the case of Afghanistan. U.S. forces bombed Serbia in 1999 and launched dozens of cruise missiles at targets in Afghanistan and the Sudan in 1998 without explicit Security Council authorization. While charges of aggression are unlikely to be brought against U.S. officials *ex post facto* for military actions in Iraq and elsewhere—certainly not for actions before July 2002 as limited by the Rome Statute—submitting to the jurisdiction of an international court that judges undefined crimes would be highly irresponsible and an open invitation to levy such charges against U.S. officials in future conflicts.

If the U.S. becomes an ICC party, every decision by the U.S. to use force, every civilian death resulting from U.S. military action and every allegedly abused detainee could conceivably give cause to America's enemies to file charges against U.S. soldiers and officials. Indeed, any U.S. "failure" to prosecute a high-ranking U.S. official in such instances would give a cause of action at the ICC. For example, the principle of complementarity will not prevent a politicized prosecutor from bringing charges against a sitting U.S. President or Secretary of Defense. That is, the U.S. Department of Justice is unlikely to file criminal charges against such officials for their decisions involving the use of military force. This decision not to prosecute would be a prerequisite for the ICC taking up the case.

At best, the U.S. would find itself defending its military and civilian officials against frivolous and politically motivated charges submitted to the ICC prosecutor. At worst, international political pressure could compel the ICC's prosecutor to file charges against current or former U.S. officials. Until the crime of aggression is defined, U.S. membership in the ICC is premature.

What the U.S. Should Do

The serious flaws that existed when President Clinton signed the Rome Statute in December 2000 continue to exist today. The Bush Administration's policy toward the ICC was prudent and in the best interests of the U.S., its officials, and particularly its armed forces. Since the ICC came into existence, the U.S. has treaded carefully by supporting the ICC on an ad hoc basis without backing away from its long-standing objections to the court. The U.S. has simultaneously taken the necessary steps to protect U.S. persons from the court's illegitimate claims of jurisdiction.

Despite intense pressure to overturn U.S. policies toward the ICC, the Obama Administration appears to appreciate the possible ramifications of joining the court. Indeed, as a candidate, Obama expressed the need to ensure that U.S. troops have "maximum protection" from politically motivated indictments by the ICC and did not openly support ratification of the Rome Statute. However, the Obama Administration has expressed less caution than either the Bush or Clinton Administrations did about the ICC. Specifically, during her confirmation hearing as Secretary of State Hillary Clinton stated:

> The President-Elect believes as I do that we should support the ICC's investigations. . . .
>
> But at the same time, we must also keep in mind that the U.S. has more troops deployed overseas than any nation. As Commander-in-Chief, the President-Elect will want to make sure they continue to have the maximum protection. . . . Whether we work toward joining or not, we will end hostility towards the ICC, and look for opportunities to encourage effective ICC action in ways that promote U.S. interests by bringing war criminals to justice.

News reports indicate that the Obama Administration is close to announcing a change in U.S. policy toward the ICC, including affirming the 2000 signature on the Rome Statute and increasing U.S. cooperation with the court. On her recent trip to Africa, Secretary of State Clinton stated that it was "a great regret but it is a fact that we are not yet a signatory [to the Rome Statute]. But we have supported the court and continue to do so."

These steps are premature if the Administration seriously wishes to provide "maximum protection" for U.S. troops. Instead, to protect U.S. military personnel and other U.S. persons and to encourage other member states to support reforms to the Rome Statute that would address U.S. concerns, the Obama Administration should:

- *Not re-sign the Rome Statute.* The Obama Administration is under pressure to "re-sign" the Rome Statute, reversing the Bush Administration's decision. In critical ways, this would be tantamount to signing a blank check. The Rome Statute is up for review by the Assembly of States Parties in 2010, and key crimes within the court's jurisdiction have yet to be defined and long-standing U.S. objections to the treaty have yet to be addressed. The Obama Administration should use the possibility of U.S. membership as an incentive to encourage the state parties to remedy the key flaws in the Rome Statute.
- *Maintain existing Article 98 agreements.* Until the Rome Statute is reformed to address all of the U.S. concerns, the Obama Administration should confirm and endorse all existing Article 98 agreements. The U.S. is militarily engaged in Iraq and Afghanistan, has troops stationed and in transit around the globe, and in all likelihood will be involved in anti-terror activities around the world for many years. Now is not the time to terminate the legal protections enjoyed by U.S. military personnel and officials deployed in foreign nations. Even if the U.S. joins the ICC at some future date, the U.S. should not terminate the Article 98 agreements because they are consistent with the Rome Statute and would serve as a useful protection if the court overreaches.
- *Establish clear objectives for changes to the Rome Statute for the 2010 review conference that would help to reduce current and potential problems posed by the ICC.* In 2010, the Assembly of States Parties is scheduled to hold the first review conference to consider amendments to the Rome Statute. A key issue on the agenda is agreeing to a definition of the crime of aggression, which is technically under the ICC's jurisdiction, but remains latent due to the states parties' inability to agree to a definition. Rather than accede to an anodyne definition, the U.S. should either seek an explicit, narrow definition to prevent politicization of this crime or, even better, seek to excise the crime from the Rome Statute entirely, on the grounds that it infringes on the Security Council's authority. Moreover, the review conference should reverse the Rome Statute's violation of customary

international law by explicitly limiting the ICC's jurisdiction only to nationals of those states that have ratified or acceded to the Rome Statute and to nationals of non-party states when the U.N. Security Council has explicitly referred a situation to the ICC.

- *Approach Security Council recommendations to the ICC on their merits and oppose those deemed detrimental to U.S. interests.* The U.S. abstentions on Security Council resolutions on Darfur indicate only that it is not U.S. policy to block all mentions of the ICC. However, accepting the reality of the ICC does not mean that the U.S. should acquiesce on substantive issues when they may directly or indirectly affect U.S. interests, U.S. troops, U.S. officials, or other U.S. nationals. Many concerns about the Rome Statute have not yet been adequately addressed. The U.S. should abstain if the resolution addresses issues critical to U.S. interests and would not directly or indirectly undermine the U.S. policy of opposing ICC claims of jurisdiction over U.S. military personnel and its nationals. Moreover, the U.S. should insist that all resolutions include language protecting military and officials from non-ICC states participating in U.N. peacekeeping operations.

Conclusion

While the International Criminal Court represents an admirable desire to hold war criminals accountable for their terrible crimes, the court is flawed notionally and operationally. The ICC has not overcome many of the problems plaguing the ad hoc tribunals established for Yugoslavia and Rwanda. It remains slow and inefficient. Worse, unlike ad hoc tribunals, it includes a drive to justify its budget and existence in perpetuity rather than simply completing a finite mission.

Its broad autonomy and jurisdiction invite politically motivated indictments. Its inflexibility can impede political resolution of problems, and its insulation from political considerations can complicate diplomatic efforts. Efforts to use the court to apply pressure to inherently political issues and supersede the foreign policy prerogatives of sovereign nations—such as the prosecutor's decision to consider Israel's actions in Gaza—undermine the court's credibility and threaten its future as a useful tool for holding accountable the perpetrators of genocide, war crimes, and crimes against humanity.

President Clinton considered the ICC's flaws serious enough to recommend against U.S. ratification of the Rome Statute unless they were resolved, and President Bush concurred. These issues remain unresolved and continue to pose serious challenges to U.S. sovereignty and its national interests. Unless the serious flaws are addressed fully, President Obama should similarly hold the ICC at arm's length. To protect its own interests and to advance the notion of a properly instituted international criminal court, the U.S. should continue to insist that it is not bound by the Rome Statute and does not recognize the ICC's authority over U.S. persons and should exercise great care when deciding to support the court's actions.

Brett Schaefer is the Jay Kingham fellow in international regulatory affairs at the Heritage Foundation. He has an MA degree in international development economics from the School of International Service at American University. He has published an edited book, *ConUNdrum: The Limits of the United Nations and the Search for Alternatives* (Rowman & Littlefield, 2009).

Steven Groves is Bernard and Barbara Lomas fellow at the Margaret Thatcher Center for Freedom, a division of the Kathryn and Shelby Cullom Davis Institute for International Studies, the Heritage Foundation. He has served as senior counsel to the U.S. Senate Permanent Subcommittee on Investigations and as an assistant attorney general for the state of Florida. Groves received his law degree from Ohio Northern University.

Michael P. Scharf

 **NO**

Is a U.N. International Criminal Court in the U.S. National Interest?

Going into the Rome Diplomatic Conference [on establishing the International Criminal Court], both the U.S. Congress and the Administration in principle recognized the need for a permanent international criminal court. Any discussion of what happened in Rome must begin by recalling the case for such an institution.

In his book, *Death by Government*, Professor Rudi Rummel, who was nominated for the Nobel Peace Prize, documented that 170 million civilians have been victims of war crimes, crimes against humanity, and genocide during the 20th Century. We have lived in a golden age of impunity, where a person stands a much better chance of being tried for taking a single life than for killing ten thousand or a million. Adolf Hitler demonstrated the price we pay for failing to bring such persons to justice. In a speech to his commanding generals on the eve of his campaign into Poland in 1939, Hitler dismissed concerns about accountability for war crimes and acts of genocide by stating, "Who after all is today speaking about the destruction of the Armenians." He was referring to the fact that the Turkish leaders were granted amnesty in the Treaty of Lausanne for the genocidal murder of one million Armenians during the First World War. After the Second World War, the international community established the Nuremberg Tribunal to prosecute the major Nazi war criminals and said "Never Again!"—meaning that it would never again sit idly by while crimes against humanity were committed. Shortly thereafter, the U.N. began work on the project to establish a permanent Nuremberg Tribunal.

But because of the cold war, the pledge of "never again" quickly became the reality of again "and again" as the world community failed to take action to bring those responsible to justice when 2 million people were butchered in Cambodia's killing fields, 30,000 disappeared in Argentina's Dirty War, 200,000 were massacred in East Timor, 750,000 were exterminated in Uganda, 100,000 Kurds were gassed in Iraq, and 75,000 peasants were slaughtered by death squads in El Salvador. Just as

Adolf Hitler pointed to the world's failure to prosecute the Turkish leaders, [Bosnian Serb leaders] Radovan Karadzic and Ratko Mladic were encouraged by the world's failure to bring [Cambodian leader] Pol Pot, [Ugandan President] Idi Amin, and [Iraqi President] Saddam Hussein to justice for their international crimes.

Then, in the summer of 1992, genocide returned to Europe just when the U.N. Security Council was freed of its cold war paralysis. Against great odds, a modern day Nuremberg Tribunal was established in The Hague to prosecute those responsible for atrocities in the Former Yugoslavia. Then a year later, genocide reared its ugly head again, this time in the small African country of Rwanda where members of the ruling Hutu tribe massacred 800,000 members of the Tutsi tribe. In the aftermath of the bloodshed, Rwanda's Prime Minister-designate (a Tutsi) pressed the Security Council: "Is it because we're Africans that a similar court has not been set up for the Rwanda genocide." The Council responded by establishing a second international war crimes Tribunal in Arusha, Tanzania.

With the creation of the Yugoslavia and Rwanda Tribunals, there was hope that ad hoc tribunals would be set up for crimes against humanity elsewhere in the world. Genocidal leaders and their followers would have reason to think twice before committing atrocities. But then something known in government circles as "Tribunal fatigue" set in. The process of reaching agreement on the tribunal's statute, electing judges, selecting a prosecutor and staff, negotiating headquarters agreements and judicial assistance pacts, and appropriating funds turned out to be too time consuming and politically exhausting for the members of the Security Council. A permanent international criminal court was universally hailed as the solution to the problems that afflict the ad hoc approach. As President [Bill] Clinton said on the eve of the Rome Conference:

> "We have an obligation to carry forward the lessons of Nuremberg. . . . Those accused of war

Senate Hearing before the Subcommittee on International Operations of the Committee on Foreign Relations, United States Senate, July 23, 1998.

crimes, crimes against humanity and genocide must be brought to justice . . . There must be peace for justice to prevail, but there must be justice when peace prevails."

So what went wrong in Rome? Why at the last minute did the United States Delegation feel compelled to join a handful of rogue States and notorious human rights violators such as Iran, Libya, China, and Iraq in voting against the statute for a Permanent International Criminal Court, while all of our allies (except Israel) voted in favor of the Court?

Rome represented a tension between the United States, which sought a Security Council-controlled Court, and most of the other countries of the world which felt no country's citizens who are accused of war crimes or genocide should be exempt from the jurisdiction of a permanent international criminal court. The justification for the American position was that, as the world's greatest military and economic power, more than any other country the United States is expected to intervene to halt humanitarian catastrophes around the world. The United States' unique position renders U.S. personnel uniquely vulnerable to the potential jurisdiction of an international criminal court. In sum, the Administration feared that an independent ICC Prosecutor would turn out to be (in the words of one U.S. official) an "international Ken Starr."

The rest of the world was in fact somewhat sympathetic to the United States' concerns. What emerged from Rome was a Court with a two-track system of jurisdiction. Track One would constitute situations referred to the Court by the Security Council. This track would create binding obligations on all states to comply with orders for evidence or the surrender of indicted persons under Chapter VII of the U.N. Charter. This track would be enforced by Security Council imposed embargoes, the freezing of assets of leaders and their supporters, and/or by authorizing the use of force. It is this track that the United States favored, and would be likely to utilize in the event of a future Bosnia or Rwanda. The second track would constitute situations referred to the Court by individual countries or the ICC Prosecutor. This track would have no built in process for enforcement, but rather would rely on the good-faith cooperation of the Parties to the Court's statute. Everyone recognized that the real power was in the first track. But the United States still demanded protection from the second track of the Court's jurisdiction. Thus, the following protective mechanisms were incorporated into the Court's Statute at the urging of the United States:

First of all, the Court's jurisdiction under the second track would be based on a concept known as "complementarity," which was defined as meaning the Court would be a last resort which comes into play only when domestic authorities are unable or unwilling to prosecute. Under this principle, for example, the Court would not have had jurisdiction over the infamous My Lai [South Vietnam] massacre [in 1968] since the United States convicted Lt. Calley [U.S. Army] and prosecuted his superior officer, Captain [Ernest] Medina.

Second, the ICC Statute specifies that the Court would have jurisdiction only over "serious" war crimes that represent a "policy." Thus, random acts of U.S. personnel, such as the downing [in 1988] of the Iran Airbus [killing 296 civilian crewmembers and passengers] by the USS *Vincennes*, would not be subject to the Court's jurisdiction.

Third, the Statute guards against spurious complaints by the ICC prosecutor by requiring the approval of a three-judge pre-trial chamber before the prosecution can launch an investigation. And the decision of the chamber is subject to interlocutory appeal to the Appeals Chamber. Fourth, the Statute allows the Security Council to affirmatively vote to postpone an investigation or case for up to twelve months, on a renewable basis. This gives the United States and the other members of the Security Council a collective (though not individual) veto over the Court where the Council is seized of a matter.

Finally, the Diplomatic Conference adopted the U.S. proposals for the selection of judges to ensure against a politicized Court. While researching my book, *Balkan Justice*, I observed the first trial before the Yugoslavia Tribunal in The Hague. I can tell you that those judges were truly independent. They did not in any way reflect the predispositions of their home countries. The selection process produced a bench made up of the most distinguished international jurists in the world. And the Yugoslavia Tribunal's jurisprudence to date reflects a respect for the rights of the defendant every bit as strong as that found in U.S. courts. The experience with the Yugoslavia Tribunal can give us comfort that a permanent international criminal tribunal would be no kangaroo court.

The United States Delegation played hardball in Rome and got just about everything it wanted. These protections proved sufficient for other major powers including the United Kingdom, France and Russia. But without what would amount to an ironclad exemption for U.S. servicemen, the United States felt compelled to force a vote, and ultimately to vote against the Court. The final vote on the Statute was 120 in favor, 7 against, with 21 abstentions. I understand that the delegates loudly cheered for fifteen minutes when the tally was announced. I'm told that a few of the members of the U.S. Delegation had tears in their eyes.

The ICC Statute will come into force when 60 countries ratify it, which given the overwhelming vote in favor, should be within a relatively short period of time. Where does that leave us? Within five years the world will have a permanent international criminal court even without U.S. support. As a non-party, the U.S. will not be bound to cooperate with the Court. But this does not guarantee complete immunity from the Court. It is important to understand that U.S. citizens, soldiers, and officials could still be indicted by the Court and even arrested and surrendered to the Court while they are visiting a foreign country which happens to be a party to the Court's Statute.

Moreover, by failing to sign the Statute, the U.S. will be prevented from participating in the preparatory committee which will draft the Court's Rules of Procedure and further define the elements of the crimes within the Court's jurisdiction. Also, by failing to sign the Statute, the U.S. will be prevented from nominating a candidate for the Court's bench, participating in the selection of the Court's Prosecutor and judges, or voting on its funding. The most important question, which cannot be answered at this time, is whether the adverse diplomatic fallout from the United States' action in Rome will ultimately prevent it from being able to utilize the first track of the Court's jurisdiction: that is, Security Council referral of cases.

The worst thing about the U.S. decision to break consensus and vote against the permanent international criminal court is that the Rome Conference will end up sending a mixed message to future war criminals and genocidal leaders. The U.S. action may be viewed as evidence that the world's greatest power does not support the international effort to bring such persons to justice. A future Adolf Hitler may point to the U.S. action in telling his followers that they need not fear being held accountable.

In the final analysis, the U.S. may have lost far more than it gained by voting against the ICC Statute. After having won so many battles in Rome, it is not clear why the U.S. Delegation did not declare victory and vote in favor of the Court (though ratification may have had to await a more favorable political climate). There's still time for a change of heart. After all, it took the United States over thirty years to ratify the 1948 Genocide Convention. But we finally did the right thing.

MICHAEL P. SCHARF is a professor of law at and director of the Center for International Law and Policy, New England School of Law. During the 1990s he served in the Office of the Legal Adviser of the U.S. Department of State, where he held the positions of attorney-adviser for law enforcement and intelligence, attorney-adviser for United Nations Affairs, and delegate to the United Nations Human Rights Commission.

EXPLORING THE ISSUE

Is U.S. Refusal to Join the International Criminal Court Justifiable?

Critical Thinking and Reflection

1. If you are opposed to the United States being subject to the ICC, what changes in the ICC's procedures and jurisdiction would change your mind?
2. What do you make of the following results of two surveys taken just a few months apart in 2002–2003? In the first survey, 71 percent of Americans said the United States should agree to the ICC as a court that could try individuals for war crimes "if their own country won't try them." The second survey asked about support of the ICC given that it could try U.S. soldiers accused of war crimes "if the United States government refuses to try them." Only 37 percent supported the ICC on this question.
3. Small, weak countries sometimes criticize the ICC because all its investigations and indictments have so far involved small, weak countries. Is this a valid criticism or does the blame properly rest with powerful countries that would block access by the ICC?

Is There Common Ground?

With the ICC treaty in effect, the countries that were a party to it met in 2003 and elected the court's 18 judges and its chief prosecutor. The following year the ICC began operations at its seat in The Hague, the Netherlands. Soon thereafter, the ICC prosecutor launched several investigations, mostly focusing on conflicts in central Africa and in the Darfur region of Sudan in northeast Africa. As of late 2012, the ICC was conducting investigations in seven countries (Central African Republic, Uganda, Kenya, Sudan, Ivory Coast, and Libya), had indicted 29 individuals, and had five in custody awaiting trial. Most significantly, the ICC has indicted President Omar al-Bashir of Sudan for 10 counts of genocide, war crimes, and crimes against humanity, and in July 2009 issued an instrumental warrant for his arrest. As of late 2012, Bashir remains in power in Sudan and for now beyond the reach of the ICC. In other activity, the court's first trial was held. The defendant, Thomas Lubanga, a former rebel leader in the Democratic Republic of the Congo (DRC), was charged with forcibly conscripting and using "child soldiers" (under age 15). In 2012 the ICC convicted Lubanga and sentenced him to 14 years in prison.

As of October 2014, 121 countries had formally agreed to the Rome Statute and joined the Assembly of State Parties that constitutes the ICC's governing board. Most of the major countries of Western Europe, Africa, and South and Central America are now parties to the ICC, as are Canada and Japan. China, Russia, India, Iran, Israel, and most of the Arab countries are among the prominent nonadherents.

The United States has also remained among the absent. It is unlikely that will change. Barack Obama has taken a more positive approach to the ICC than President Bush did but still has made no move to ask the Senate to ratify the Treaty of Rome. Even if Obama did so, ratification would be unlikely without, at the least, much greater protections for U.S. troops abroad against possible ICC indictments.

Create Central

www.mhhe.com/createcentral

Additional Resources

Chapman, Terrence L., and Chaudoin, Stephen. (2013). "Ratification Patterns and the International Criminal Court." *International Studies Quarterly* (57/2: 400–409)

Cryer, Robert, Friman, Håkan, Robinson, Darryl, and Wilmshurst, Elizabeth. (2014). *An Introduction to International Criminal Law and Procedure* (Cambridge University Press)

Luban, David. (2013) "After the Honeymoon: Reflections on the Current State of International Criminal Justice." *Journal of International Criminal Justice* (11/3: 505–515)

Schabas, William A. (2011). *An Introduction to the International Criminal Court* (Cambridge University Press)

Internet References . . .

Coalition for the International Criminal Court

www.iccnow.org/

ICC—International Criminal Court

www.icc-cpi.int/en_menus/icc/Pages/default.aspx

Journal of International Criminal Justice

This journal is published quarterly and is an excellent source of information on the ICC.

http://jicj.oxfordjournals.org/

IntlCriminalCourt—YouTube

www.youtube.com/user/IntlCriminalCourt

Selected, Edited, and with Issue Framing Material by:
John T. Rourke, *University of Connecticut, Storrs*

ISSUE

Should the United States Ratify the Convention to Eliminate All Forms of Discrimination Against Women?

YES: **Melanne Verveer,** from Testimony During Hearings on "Ratify the CEDAW," before the Subcommittee on Human Rights and the Law, the Committee on the Judiciary, U.S. Senate (November 18, 2010)

NO: **Steven Groves,** from Testimony During Hearings on "Reject CEDAW," before the Subcommittee on Human Rights and the Law, the Committee on the Judiciary, U.S. Senate (November 18, 2010)

Learning Outcomes

After reading this issue, you will be able to:

- Recall the background of Convention to Eliminate All Forms of Discrimination Against Women (CEDAW).
- Relate how CEDAW has been dealt with by most countries and how that differs from the U.S. reception.
- Recount the reasons why some Americans oppose Senate ratification of CEDAW.
- Enumerate the counterarguments for ratifying CEDAW.

ISSUE SUMMARY

YES: Melanne Verveer, ambassador-at-large, Office of Global Women's Issues, U.S. Department of State, tells a congressional committee that the U.S. Senate should ratify the Convention on the Elimination of All Forms of Discrimination Against Women (CEDAW) because doing so would send a powerful message about the U.S. commitment to equality for women across the globe.

NO: Steven Groves, the Bernard and Barbara Lomas Fellow in the Margaret Thatcher Center for Freedom, a division of the Kathryn and Shelby Cullom Davis Institute for International Studies at the Heritage Foundation, headquartered in Washington, DC, contends that ratifying CEDAW would neither advance U.S. international interests nor enhance the rights of women in the United States.

Females constitute about half the world population, but they are a distinct economic–political–social minority because of the wide gap in societal power and resources between women and men. Women constitute 70 percent of the world's poor and two-thirds of its literates. They occupy only 14 percent of the managerial jobs, are less than 40 percent of the world's professional and technical workers, and garner only 35 percent of the earned income

in the world. Women are also disadvantaged politically. In late 2012, only 14 women were serving as the top political leader in their countries; women make up just 8 percent of all national cabinet officers; and only one of every five national legislators is a woman. On average, life for women is not only harder and more poorly compensated than it is for men, but also more dangerous. "The most painful devaluation of women," the United Nations reports, "is the physical and psychological violence that stalks them from cradle to grave." Signs of violence against women include the fact that about 80 percent of the world's refugees are women and their children. Other assaults on women arguably constitute a form of genocide. According to the UN Children's Fund, "In many countries, boys get better care and better food than girls. As a result, an estimated one million girls die each year because they were born female." None of these economic, social, and political inequities are new. Indeed, the global pattern of discrimination against women is ancient. What is new is the global effort to recognize the abuses that occur and to ameliorate and someday end them. To help accomplish that goal, the U.N. General Assembly in 1979 voted by 130 to 0 to put the Convention on the Elimination of All Forms of Discrimination Against Women (CEDAW) before the world's countries for adoption. Supporters hailed the treaty as a path-breaking step on behalf of advancing the status of women. Many countries agreed, and by September 1981 enough countries had signed and ratified CEDAW to put it into effect. This set a record for the speed with which any human rights convention had gone into force.

CEDAW is a women's international bill of rights. Most of these rights are enumerated in various other treaties as applicable to all humans, but women's rights had not been specifically and fully addressed in any other treaty before CEDAW. Countries that legally adhere to the convention agree to undertake measures to end all the various forms of discrimination against women. Doing so entails accepting legal gender equality and ensuring the practice of gender equality by abolishing all discriminatory laws, enacting laws that prohibit discrimination against women, and establishing agencies to protect women's rights.

President Jimmy Carter signed CEDAW in 1980 and submitted it to the Senate for ratification. However, he was soon thereafter defeated for reelection, and the treaty languished in legislative limbo ever since. President Bill Clinton did make an effort to move CEDAW forward, but he was unsuccessful. Hope among those who support ratifying CEDAW rose anew when President Barack Obama was elected president and he appointed Hillary Clinton as secretary of state. Both supported CEDAW. Additionally, the Democrats controlled Congress. This led to hearings before a Senate committee in 2010. The testimony that constitutes the YES and NO selections comes from those hearings. In the YES selection, Melanne Verveer urges the adoption of CEDAW and tells the Senate that it is long overdue for the United States to stand with the women of the world in their effort to obtain the basic rights that women in this country enjoy. Steven Groves disagrees, telling the committee that U.S. ratification of CEDAW would have little or no impact on women's rights globally and that it would unnecessarily and unwisely subject the United States to interference by the United Nations and international courts.

YES ⬅

Melanne Verveer

Ratify the CEDAW

Thank you for this opportunity to discuss with you the United Nations Convention on the Elimination of All Forms of Discrimination Against Women, commonly known as CEDAW, or the Women's Treaty. . . . Today, I would like to talk about what the Women's Treaty represents and why U.S. ratification is critical to our efforts to promote and defend the rights of women across the globe.

This hearing could not come at a more critical time for the world's women. Gender inequality and oppression of women are rampant across the globe. The scale and savagery of human rights violations committed against women and girls is nothing short of a humanitarian tragedy. Today, violence against women is a global pandemic. In some parts of the world, such as the Democratic Republic of Congo, Burma, and Sudan, women are attacked as part of a deliberate and coordinated strategy of armed conflict where rape is used as a tool of war. In others, like Afghanistan, girls are attacked with acid and disfigured simply because they dare attend school. Girl infanticide and neglect has contributed to the absence from school of an estimated 100 million girls worldwide. In places where girls are not as valued and there is a strong preference for sons, practices ranging from female genital mutilation, to child marriage, to so-called "honor killings," to the trafficking of women and girls into modern-day slavery highlight the low status of females around the globe.

In far too many places, women's participation in parliaments, village councils and peace negotiations is circumscribed or prevented altogether. Policies instructing that "women need not apply" continue to limit employment opportunities and pay. The majority of the world's illiterate are women and, according to the World Bank, girls constitute 55 percent of all out-of-school children. This has devastating consequences on the health and well-being of families and communities. And today, the HIV-AIDS pandemic has a woman's face, with the number of infections rising at alarming rates among adolescent girls in many places who face the threat of violence, including sexual violence, in their lives.

Women's equality has rightly been called the moral imperative of the 21st century. Where women cannot participate fully and equally in their societies, democracy is a contradiction in terms, economic prosperity is hampered, and stability is at risk. Standing up against the appalling violations of women's human rights around the globe, and standing with the women of the world, is what ratifying the Women's Treaty is about.

Why the United States Should Ratify the Women's Treaty

In my time at the State Department, I have visited scores of countries and met with women from all walks of life, from human rights activists in Russia, to microcredit recipients and small-business entrepreneurs in rural South Asia, to survivors of rape and conflict in the Democratic Republic of the Congo. In my travels, the number-one question I am asked time and time again is, "Why hasn't the United States ratified CEDAW?"

It is understandable that I continue to receive this question everywhere I go. The United States has long stood for the principles of equal justice, the rule of law, respect for women, and the defense of human dignity. We know that women around the world look to the United States as a moral leader on human rights. And yet when it comes to the Women's Treaty, which reflects the fundamental principle that women's rights are human rights, we stand with only a handful of countries that have not ratified, including Somalia, Iran, and Sudan—countries with some of the worst human rights records in the world. We stand alone as the only industrialized democracy in the world that has not ratified the Women's Treaty. And we stand on the sidelines, unable to use the Women's Treaty to join with champions of human rights who seek to use it as a means to protect and defend women's basic human rights.

U.S. Senate, November 18, 2010.

U.S. ratification of the Women's Treaty matters because the moral leadership of our country on human rights matters. Some governments use the fact that the U.S. has not ratified the treaty as a pretext for not living up to their own obligations under it. Our failure to ratify also deprives us of a powerful tool to combat discrimination against women around the world, because as a non-party, it makes it more difficult for us to press other parties to live up to their commitments under the treaty.

The United States is firmly committed to the principles of women's equality as enshrined in the U.S. Constitution. Our ratification will send a powerful and unequivocal message about our commitment to equality for women across the globe. It will lend much needed validation and support to advocates fighting the brutal oppression of women and girls everywhere, who seek to replicate in their own countries the strong protections against discrimination that we have in the United States. And it will signal that the United States stands with the women of the world. Importantly, ratification will also advance U.S. foreign policy and national security interests. As the Obama Administration has made clear, women's equality is critical to our national security. President Obama's National Security Strategy recognizes that "countries are more peaceful and prosperous when women are accorded full and equal rights and opportunity. When those rights and opportunities are denied, countries lag behind." And as Secretary [Hillary] Clinton has stated, "the subjugation of women is a threat to the national security of the United States. It is also a threat to the common security of our world, because the suffering and denial of the rights of women and the instability of nations go hand in hand." Ratification of this treaty, which enshrines the rights of women in international law, is not only in the interest of oppressed women around the world—it is in our interest as well.

In fact, my office has been working closely with the Office of the Under Secretary of Defense for Policy at the Department of Defense to highlight issues related to women, peace and security. We as a U.S. government recognize the interconnection of women's progress and the advancement of U.S. objectives across the world. And Admiral Mullen, Chairman of the Joint Chiefs of Staff, recently stated, "Secretary of State Hillary Clinton wisely summed it up last week when she said, 'If we want to make progress towards settling the world's most intractable conflicts, let's enlist women.' I couldn't agree more—and I would only add: The time to act is now so we don't have to ask, yet again, why did this take so long? But as we think about how far we've come, we must also consider how far we have still to go."

How the Women's Treaty Helps Eliminate Discrimination Against Women

I would like to briefly describe what the Women's Treaty is, the principles it enshrines, and how it can be used to challenge discrimination against women around the world. The Women's Treaty was adopted by the United Nations nearly 31 years ago and is the first treaty to comprehensively address women's rights and fundamental freedoms. The treaty builds on several previous international human rights instruments, including the UN Universal Declaration of Human Rights, and the International Covenant on Civil and Political Rights (ICCPR). It obliges parties to end discrimination against women and addresses areas that are crucial to women's equality, from citizenship rights and political participation to inheritance and property rights to freedom from domestic violence and sex trafficking. It is consistent with the approach that we have already taken on these issues domestically. To date, 186 out of 192 UN member states are party to the treaty.

Around the world, women are using the Women's Treaty as an instrument for progress and empowerment. There are countless stories of women who have used their countries' commitments to the treaty to bring constitutions, laws, and policies in line with the principle of nondiscrimination against women. Over the course of my travels, I have seen firsthand its incredible influence in helping women change their societies. Today, I would like to highlight just a few examples that illuminate the treaty's ability to help women push for equal treatment in their communities.

Morocco: The Women's Treaty has been used to fight discrimination against women in family law. For example, in Morocco, for nearly a century, family law was largely determined by differing interpretations of Islamic law, which resulted in oppression and unequal treatment for wives. Brides were not asked to give their consent to marriage during the wedding ceremony. Polygamy was widespread, and husbands had the power to "repudiate" a marriage without court proceedings or their wives' consent. Women in Moroccan civil society worked tirelessly and even faced imprisonment in their effort to end discrimination against women in family law, but they did not back down. In 1993, Morocco ratified the Women's Treaty with a set of reservations, and in 2004, a new Morocco Family Code was enacted that protected women's rights in matters of marriage and family relations. Today, women no longer need a matrimonial guardian to determine whom they will marry. In addition, a woman can now initiate divorce proceedings, which are now determined in a court of law, and there are a series of restrictions in place making polygamy far more difficult to practice.

Afghanistan: The Women's Treaty has also been used to combat discrimination against women even in countries that fall far short of their commitment to women's equality under the treaty, such as Afghanistan. As we know, under the brutal Taliban regime, Afghan women and girls suffered untold deprivations of their basic human rights, including the right to attend school, thereby penalizing an entire generation. The fact that Afghanistan is party to agreements like the ICCPR and the Women's Treaty has helped to provide legitimacy for women's rights advocates seeking to improve conditions for women and girls. Indeed, Afghan activists recently pushed for a new law to eliminate violence against women. And several Afghan women's organizations have banded together to release their own "shadow report" detailing the government's actions to prevent and respond to violence against women. Thanks to the efforts of women's advocates, the Afghan government—for the first time since ratifying the Women's Treaty—is working to prepare a public report on its implementation of the treaty.

Mexico: The Women's Treaty has also been used to combat violence against women and sexual assault. In Mexico, for example, the treaty was deployed as a tool against violence in some of the country's most dangerous areas. An estimated 450 girls and women have been killed in Ciudad Juárez and Chihuahua City since 1993. According to Mexican authorities, most of these women were sexually assaulted before their murders. Local human rights groups report that few cases have been investigated and in even fewer have perpetrators been brought to justice. But in 2007, human rights groups won a major victory with the enactment of a national law inspired, in large part, by the Women's Treaty. The new Mexican law requires federal, state and local authorities to coordinate activities to prevent and respond to violence against women, and authorizes the Interior Minister to declare a state of alert if he or she determines there is an outbreak of widespread gender-based violence.

Philippines: The Women's Treaty has provided activists around the world with a useful framework for women's human rights that has advanced and improved laws prohibiting discrimination against women. For instance, in the Philippines, the treaty was heavily relied upon as a blueprint for framing the first Magna Carta of Women, a comprehensive equal-rights statute that provides political, civil, and economic rights for all Filipino women, with special protections for those who are members of marginalized groups. Women's groups, working in coordination with international organizations, used the Women's Treaty to help develop a definition of gender discrimination and outline the responsibilities of the government to protect its citizens. This historic and far-reaching law was signed into law by President Gloria Arroyo in 2009. Among its several provisions, the Magna Carta affirms Filipino women's rights to education, political participation and representation, and equal treatment before the law.

Uganda: The Women's Treaty has also been used to achieve equal treatment for women in the critical area of land rights. In some parts of the world, women produce 70 percent of the food and yet earn only 10 percent of the income and own only 1 percent of the land—a situation that is not only unfair, but also relegates women to lives of poverty. In Uganda, a robust women's movement has made efforts to tackle this problem by relying on both the Women's Treaty and national legislation to pursue land ownership rights and challenge customary land tenure practices. Empowered by the Women's Treaty and the enactment of the country's Land Act in 1998, women's groups and activists began a tireless campaign to ensure that women were protected in the tenure, ownership and administration of land. In their fight for equal treatment, these activists continue to rely on the Women's Treaty.

Conclusion

Fifteen years ago, as First Lady of the United States, Hillary Clinton addressed the UN Fourth World Conference on Women in Beijing and proclaimed that women's rights are human rights. Today, the litany of abuses against women that she described in her address—from violence against women to trafficking to female genital mutilation to girl infanticide—persists. We cannot stand by while girls and women continue to be fed less, fed last, overworked, underpaid, subjected to violence both in and out of their homes—in short, while discrimination against women and girls remains commonplace around the globe. For as long as the oppression of women continues, the peaceful, prosperous world we all seek will not be realized. It has been over 30 years since the Women's Treaty was first adopted by the United Nations. Since that time, as I have described, the treaty has been used to advocate for and realize equal treatment for women and girls around the world. But much work remains to be done. And it is long overdue for the United States to stand with the women of the world in their effort to obtain the basic rights that women in this country enjoy.

As Secretary Clinton has said, "the United States must remain an unambiguous and unequivocal voice in support of women's rights in every country, every region, on every continent." By ratifying the Women's Treaty, we will speak with this clarity of voice and purpose. We will strengthen the efforts of those who toil for women's rights, for equal treatment, and for human dignity. And we will make clear our belief that human rights are women's rights and women's rights are human rights, once and for all.

MELANNE VERVEER is ambassador-at-large, Office of Global Women's Issues, U.S. Department of State. She earlier served as chief of staff for first lady Hillary Rodham Clinton and was chairman of the board of the Vital Voices Global Partnership, an international private organization supporting global women's leadership. She has an MA from Georgetown University.

Steven Groves

 **NO**

Reject CEDAW

Thank you for inviting me to testify before you today regarding the Convention on the Elimination of All Forms of Discrimination Against Women (CEDAW). Ratification of CEDAW would neither advance U.S. national interests within the international community nor enhance the rights of women in the United States. Domestically, CEDAW membership would not improve our existing comprehensive statutory framework or strengthen our enforcement system for the protection of women's rights.

Within the international sphere, the United States need not become party to the Convention to demonstrate to the rest of the world its commitment to women's rights at home or abroad. Becoming a member of CEDAW would produce, at best, an intangible and dubious public diplomacy benefit.

Moreover, it does not serve the interests of the United States to periodically submit its record on women's rights to scrutiny by a committee of gender experts that has established a record of promoting policies that do not comport with existing American norms and that encourages national governments to engage in social engineering on a massive scale.

The United States should become party to a treaty only if membership would advance U.S. national interests. For a human rights treaty such as CEDAW, national interests may be characterized in both domestic and international terms. Only if U.S. membership in CEDAW would advance the cause of women's rights domestically and further U.S. national interests in the world should the United States consider ratification of the treaty.

Domestically, ratification of CEDAW is not needed to end gender discrimination or advance women's rights. The United States already has effective avenues of enforcement in place to combat discrimination based on sex. Specifically, in addition to the Equal Protection Clause of the Fourteenth Amendment to the U.S. Constitution, the United States has in place a wide range of state and federal laws to protect and advance women's rights

concerning their employment, compensation, housing, education, and other areas. Federal laws include, but are not limited to:

- Title VII of the Civil Rights Act of 1964, which prohibits discrimination in employment on the basis of, *inter alia*, sex and has been interpreted to prohibit sexual harassment or the creation of a hostile working environment;
- The Pregnancy Discrimination Act, enacted in 1978, which prohibits discrimination on the basis of pregnancy and childbirth;
- The Equal Pay Act of 1963, which prohibits discrimination on the basis of sex in regard to the compensation paid to men and women for substantially equal work performed in the same establishment;
- The Fair Housing Act of 1968, which prohibits discrimination in the sale or rental of housing on the basis of, *inter alia*, sex;
- Title IX of the Education Amendments of 1972, which prohibits discrimination on the basis of sex in federally funded education programs or activities;
- The Equal Credit Opportunity Act, enacted in 1974, which prohibits discrimination against credit applicants on the basis of, *inter alia*, sex;
- The Violence Against Women Act of 1994, which was intended to improve criminal justice and community responses to acts of domestic violence, dating violence, sexual assault, and stalking; and
- The Lilly Ledbetter Fair Pay Act of 2009, which revised the statute of limitations requirements in equal-pay lawsuits to assist women in recovering wages lost due to discrimination.

Taking this extensive legal framework into consideration, it is difficult to imagine how membership in CEDAW will further advance the protections provided to women in the United States. In fact, the protections provided by the U.S. Constitution and existing U.S. law exceed the

U.S. Senate, November 18, 2010.

provisions in the treaty. This legal framework serves as a foundation that can be modified or expanded as necessary through the democratic process.

Some of this federal legislation remains controversial and will continue to be debated in Congress and litigated in U.S. courts. Differences of opinion regarding these laws and the extent of constitutional protections based on gender are likely to persist for years to come. The resolution of these issues should be sought through domestic legislative and judicial avenues rather than through the judgment of gender experts sitting on the CEDAW Committee who may possess inadequate specific knowledge or understanding of U.S. laws and practices. Robust debate regarding these issues continues in the United States despite the fact that it is not a party to CEDAW. For instance, the existence of pending federal legislation on the "pay equity" issue— such as the Paycheck Fairness Act—indicates an ongoing effort in Congress to resolve gender and compensation issues through the traditional democratic process.

Unlike the expansive provisions of CEDAW and the overly broad recommendations of the CEDAW Committee, these federal laws were crafted to address specific issues of gender discrimination in the United States, not to address the general policy opinions of the international community.

Measuring whether U.S. membership in CEDAW would actually improve the image of the United States abroad may be impossible.

Those who say that ratification would allow the United States to claim the moral high ground within the international community—at least in regard to women's rights—imply that the United States is deficient in protecting those rights, when in truth the United States has been a leader and standard bearer for empowering women. It already holds the moral high ground. Ratifying a treaty merely to score points overseas is not a sound justification for a decision that could have unforeseen or negative domestic ramifications.

The United States amply demonstrated to the international community that it is committed to the protection of women's rights when, in 1992, it ratified the International Covenant on Civil and Political Rights (ICCPR). By ratifying the ICCPR, the United States made an international political commitment to guarantee the individual liberties, political rights, and physical integrity of its citizens "without distinction of any kind, such as race, colour, sex, language, religion, political or other opinion, national or social origin, property, birth or other status." Indeed, Article 3 of the ICCPR requires that the United States "undertake to ensure the equal right of men and women

to the enjoyment of all civil and political rights set forth in the present Covenant."

Moreover, the United States has demonstrated in the past and continues to demonstrate its commitment to women's rights not just in the U.S., but around the world as well, regardless of the fact that it is not a member of CEDAW. The presence at this very hearing of Ambassador Melanne Verveer, the first Ambassador-at-Large for Global Women's Issues, is only the latest indication that the U.S. is committed to the political, economic, and social empowerment of women around the globe. Secretary Clinton's establishment of the International Fund for Women further indicates that the U.S. government is continually formulating programs for the advancement of women's causes outside of the United States.

While everyone may not agree with the policy aims of these programs, these acts by the United States are far more relevant to women in need around the world than U.S. membership in an international convention.

In short, to assert that the United States lacks credibility on the issue of international women's rights due to its non-membership in CEDAW is simply specious.

Beyond the dubious public diplomacy benefit that would allegedly be enjoyed by the United States upon ratification of CEDAW, it is difficult to determine how U.S. national interests would otherwise be advanced by participating in the central activity required by the treaty— reporting to the CEDAW Committee every four years regarding the U.S. record on women's rights. The CEDAW Committee has for 30 years established a consistent record of promoting gender-related policies that do not comport with existing American legal and cultural norms and has encouraged the national governments of CEDAW members to engage in social engineering on a massive scale.

For instance, Article 5 of CEDAW compels members of the treaty to "modify the social and cultural patterns of conduct of men and women, with a view to achieving the elimination of prejudices and customary and all other practices which are based on . . . stereotyped roles for men and women." The CEDAW Committee has cited this provision over the years to oblige member states to seek the modification of the roles of men and women as husbands, wives, caregivers, and breadwinners.

The Committee appears to be particularly contemptuous of the role of women as mothers and caregivers. For example, in 1999 it determined that "the persistence of the emphasis on the role of women as mothers and caregivers [in Ireland] tends to perpetuate sex role stereotypes and constitutes a serious impediment to the full implementation of the Convention." In its report on Georgia, the

Committee expressed concern over "the stereotyped roles of women . . . based on patterns of behaviour and attitudes that overemphasize the role of women as mothers." In 2000, the Committee issued its now famous concluding observation to Belarus in which it referred to Mothers' Day as a stereotypical symbol and chided Belarus for "encouraging women's traditional roles."

The CEDAW Committee has made other policy choices that are inconsistent with U.S. societal norms. Prostitution, in particular, has been treated by the Committee not as a crime that should be discouraged, but rather as a reality that should be tolerated and regulated. Indeed, the Committee appears to have little or no regard for the moral choices made by member states concerning whether they consider prostitution to be a criminal act that should be prohibited. For instance, in 2001, a representative from Guinea told the Committee that prostitution was "one of the social scourges" in Guinea and was illegal and "rejected and condemned by society." Undaunted, the Committee ignored Guinea's social and cultural norms and urged the government not to "penaliz[e] women who provide sexual services." Similarly, in 1999, the Committee instructed Liechtenstein to "review . . . the law relating to prostitution to ensure that prostitutes are not penalized." Also in 1999, the Committee told China that it was "concerned that prostitution . . . is illegal in China." Rather than recommending that China take steps to reduce poverty and enhance economic freedom to alleviate the problem, the Committee "recommend[ed] decriminalization of prostitution."

Many of the policy choices prescribed by the CEDAW Committee are at odds with other social, political, cultural, legal, and democratic norms in the United States. The Committee supports the concept of "comparable worth" to address allegations of gender discrimination in compensation. It advocates the use of quota systems to achieve de facto equality in various fields including education, politics, and employment—a policy that demands equal outcomes rather than equal opportunity. Finally, and controversially, the Committee regularly instructs member states to amend their laws to ease restrictions on abortion.

In short, it does not advance U.S. national interests to submit itself to scrutiny every four years by a committee of gender experts that has already demonstrated its divergence with U.S. policy choices, including on highly controversial issues regarding American social and cultural norms. In the eyes of the CEDAW Committee, the current laws and norms of the United States place it in direct and flagrant violation of CEDAW's provisions, even though those provisions arguably seek to dictate women's roles in a manner that may not be welcomed by all women or, in the case of prostitution, may be antithetical to their welfare.

The United States should not make international political commitments that it cannot keep due to its own legal, social, and cultural traditions, and joining CEDAW will unfairly place it in an untenable position.

An objective analysis of CEDAW indicates that ratifying it would not advance U.S. national interests either at home or abroad.

U.S. ratification of CEDAW would produce, at best, an intangible and fleeting public diplomacy benefit in the international community. The United States need not become party to the convention to demonstrate its commitment to women's rights or to advance the cause of women in other nations. Any nation that questions U.S. dedication to protecting the rights of American women need only review the architecture of our laws and the network of state and federal agencies that enforce those laws.

Instead of seeking membership in CEDAW, the U.S. Congress and American civil society should continually review the implementation of existing laws barring gender discrimination in all spheres of domestic life. Those entities are far better positioned to conduct such a review than a committee of supposed gender experts from 23 foreign nations.

Steven Groves is Bernard and Barbara Lomas Fellow in the Margaret Thatcher Center for Freedom, a division of the Kathryn and Shelby Cullom Davis Institute for International Studies, The Heritage Foundation. He has served as senior counsel to the U.S. Senate Permanent Subcommittee on Investigations and as an assistant attorney general for the state of Florida. Groves received his law degree from Ohio Northern University.

EXPLORING THE ISSUE

Should the United States Ratify the Convention to Eliminate All Forms of Discrimination Against Women?

Critical Thinking and Reflection

1. Start by reading the text of CEDAW at www.un.org/womenwatch/ daw/cedaw/cedaw.htm. Then ask yourself whether it protects any rights that American women should not have. If it does so, which ones?
2. Those who worry about U.S. sovereignty argue that if U.S. law and the U.S. courts refused to uphold the rights women have under CEDAW, the women could appeal to international courts for justice. Is that ability to appeal good or bad?
3. Is U.S. ratification of CEDAW superfluous given the extensive range of rights American women already have?

Is There Common Ground?

The concentrated effort to promote women's rights internationally within the context of advancing globalism dates back only to 1975, which the UN declared the International Women's Year. There have been many changes that benefit women since that time, but those changes have only begun to ease the problems that advocates of women's rights argue need to be addressed. CEDAW has been a keystone of the international effort to promote women's rights. As of late 2014, 188 countries had adhered to CEDAW, leaving only eight countries [the United States, the Holy See (Vatican), Iran, Sudan, Somalia, Palau, newly formed South Sudan, and Tonga] that have not formally adhered to CEDAW.

The hearings held by the Senate Judiciary Committee in 2010 were more a symbolic show of support than portending any real chance the Senate would ratify the treaty. Indicative of that it is the Senate Foreign Relations Committee that has jurisdiction over treaties and would under most circumstances have to hold hearings prior to a treaty moving for debate and decision on the Senate floor. That did not happen in 2010 at least in part, analysts conclude, because the Obama administration, while supporting CEDAW, has not made a concerted effort to get the Senate to ratify it.

Create Central

www.mhhe.com/createcentral

Additional Resources

Baldez, Lisa. (2013). "The UN Convention to Eliminate All Forms of Discrimination Against Women (CEDAW): A New Way to Measure Women's Interests." *Politics & Gender* (7/3: 419–423)

Cusack, Simone. (2013). "The UN Convention on the Elimination of All Forms of Discrimination against Women: A Commentary." *Human Rights Quarterly* (35/1: 251–258)

Englehart, Neil A., and Miller, Melissa K. (2013). "The CEDAW Effect: International Law's Impact on Women's Rights." *Journal of Human Rights* (13/1: 22–47)

Hellum, Anne, and Aasen, Henriette Sinding, eds. (2013). *Women's Human Rights: CEDAW in International, Regional and National Law*, Vol. 3 (Cambridge University Press)

Internet References . . .

CEDAW 2014

http://www.womenstreaty.org/index.php
/about-cedaw/faq

UN Women—CEDAW

http://www.un.org/womenwatch/daw/cedaw
/states.htm

The Committee on the Elimination of Discrimination Against Women

http://www.ohchr.org/en/hrbodies/cedaw/pages
/cedawindex.aspx

Concerned Women for America Against CEDAW

http://www.cwfa.org/why-the-u-s-has-not-and-should
-not-ratify-cedaw/

Unit 6

Environmental and Social Issues

A key aspect of international relations over the last two centuries or so has been the "coming together" of the world, with countries, their companies and other organizations, and their people interacting across national borders at a vastly accelerated pace and on an expanding range of matters. The distinction between national and international problems has become increasingly blurred.

Even within this relatively recent phenomenon of seeing many problems as global and needing global solutions, environmental and social issues are something of the "new kids on the block." On the social front, human rights were virtually a nonissue among countries a hundred years ago, and it's even more recent that poverty, disease, and other social issues have come onto the international agenda. A central advance came in 1948 with the UN's Universal Declaration of Human Rights (UDHR), which detailed a broad range of basic civil, political, economic, social, and cultural rights that all humans should possess. In addition to such commonly thought of civil and political rights, such as freedom of religion and freedom from cruel and unusual punishment, the UDHR endorsed a range of social and economic rights. For example, Article 25 declares:

> Everyone has the right to a standard of living adequate for the health and well-being of himself and of his family, including food, clothing, housing and medical care and necessary social services, and the right to security in the event of unemployment, sickness, disability, widowhood, old age or other lack of livelihood in circumstances beyond his control.

From this beginning, the idea has slowly grown that not just individual governments for their own people, but people and countries everywhere should take steps to end such age-old oppressions as poverty, illiteracy, lack of sanitation, and inadequate health care that limit the human potential.

The environment is an even more recent entry on the global agenda. The depletion of the ozone layer, the pollution of the oceans, the decline in water resources, the disappearance of the world's forests, and global warming are but some of the problems that can only be fully addressed by agreeing there is a global responsibility to cooperate to ease or end these threats.

This unit also addresses the role of the individual in the international system. Each of us is a citizen of our country, but it is arguable that each of us is also a citizen of the world. What if the two citizenships clash? Some would argue that there is no choice—my country, right or wrong—but is that an old thought for an old world system that no longer exists?

Selected, Edited, and with Issue Framing Material by:
John T. Rourke, *University of Connecticut, Storrs*

ISSUE

Is President Obama's U.S. Global Warming Policy Wise?

YES: Barack H. Obama, from "Remarks at the League of Conservation Voters Capital Dinner," *Speeches and Remarks* (2014)

NO: James Inhofe, from "Speech on the Floor of the U.S. Senate," *Congressional Record* (2014)

Learning Outcomes

After reading this issue, you will be able to:

- Indicate what global warming is and why it alarms President Obama and most other observers.
- Discuss U.S. CO_2 emissions compared with other countries' emissions in the past, present, and in the foreseeable future.
- Distinguish between the evidence that global warming is occurring and the evidence that human activity is behind all, some share, or any of global warming.
- Understand the varying claims of President Obama and Senator Inhofe about the economic impact of the president's program to curb U.S. carbon dioxide emissions.
- Evaluate preliminarily how big a cut in U.S. emissions will be necessary to begin to reduce, or even stabilize, global emissions given the rapidly growing emissions of China, India, and other countries.

ISSUE SUMMARY

YES: Barack H. Obama, president of the United States, tells the audience at the League of Conservation Voters dinner that dealing with the rapidly growing threat of climate change is increasingly urgent because we know more about the threat we did back then and because we know through experience that we can act in ways that protect our environment and promote economic growth at the same time.

NO: James Inhofe, a Republican U.S. senator from Oklahoma and member of its Committee on Environment and Public Works, tells members of the Senate his opposition to the Obama administration's global warming policies is based on two factors: (1) the administration is intentionally ignoring the most recent science around global warming, and (2) global warming policies costing between $300 billion and $400 billion a year, along with the rest of the U.S. Environmental Protection Agency's regulations, are resulting in millions of American job losses.

The past few centuries have witnessed rapid increases in human prosperity driven in substantial part by industrialization, electrification, the burgeoning number of private and commercial vehicles, and a host of other inventions and improvements that, in order to work, consume massive amounts of fossil fuel (mostly coal, petroleum, and natural gas). The burning of fossil fuels gives off carbon dioxide (CO_2) into the atmosphere. The discharge of CO_2 from burning wood, animals exhaling, and some other sources is nearly as old as Earth itself, but the industrial and population advances since the beginning of the industrial revolution in the mid-1700s have rapidly increased the level of discharge. Since 1950 alone, annual global CO_2

emissions have more than tripled. Much of this is retained in the atmosphere because the ability of nature to cleanse the atmosphere of the CO_2 through plant photosynthesis has been overwhelmed by the vast increases in fossil fuel burning and the simultaneous cutting of vast areas of the world's forests for habitation and agriculture.

Most scientists take the view that human-generated CO_2 has at least accelerated global warming. The reason, they contend, is the greenhouse effect. As CO_2 accumulates in the upper atmosphere, it creates a blanket effect, trapping heat and preventing the nightly cooling of the Earth. Other gases, such as methane, also contribute to creating the thermal blanket. How much human-generated CO_2 is responsible to global warming is less certain.

Some believe that the current warming would be underway, at least to some degree, regardless of human activity. Over the Earth's history during the last 400 million years, there have been cycles of rising and declining temperature, each of which has lasted 150,000 years or so. The last high peak was about 125,000 years ago; the following low was about 10,000 years ago, after which temperatures began to rise again. Some argue that the current global warming trend is partly an extension of this long-term trend acceleration.

What is clear is that over the last century Earth's average temperature has risen about 1.1 degree Fahrenheit and that the last decade has been the warmest since temperature records were first kept in 1856. Most scientists also believe that global warming will have a dramatic and in some case catastrophic impact on rainfall, wind currents, and other climatic patterns. Among other impacts, the polar ice caps are already melting more quickly. As a result, sea levels will rise, displacing perhaps over 100 million people on the continents' coasts during the coming century. Some weather experts also project an increase in the number and intensity of hurricanes and other catastrophic weather events. Droughts will also dry once fertile lands. It may also be the case, though, that warming will turn once marginal lands in the north into more productive areas, and changing rain patterns may turn once arid land into fertile fields.

Even more difficult than analyzing the problem is coming up with a solution. A first international effort to control CO_2 emissions was the UN-sponsored treaty called the Kyoto Protocol in 1997. It requires the industrialized countries to significantly cut their CO_2 emissions but imposed no limits on developing countries including China and India. Objecting to this, the United States did not ratify the treaty. With the limits set by the Kyoto Protocol extending only through 2012, an international conference gathered in 2009, in Copenhagen, Denmark, to try to establish a new round of reductions. The resulting Copenhagen Accord adopted new limits, but was still

adjudged a failure by many analysts because none of the limits was binding and even the nonbinding limits were cuts to projected future levels, not from current levels. It is clear that unless countries get much more serious about reducing CO_2 emissions they will continue to grow. Indeed, even if the Copenhagen targets are met, they will only slow down the growth. Between 2000 and 2012, the world growth rate has averaged almost 3 percent annually.

Looking at emissions around the world, the pattern has changed considerably in the past two decades. Up to then, most of the emissions had come from the United States and Europe, thus arguably giving them then and now a special burden to curb global warming. More recently, though, it is the less developed countries, especially those in Asia, that have been the primary source of increased emissions. This has occurred because of the industrialization and modernization of these countries, particularly China and India. China now leads all countries in CO_2 emissions, with an output more than a third larger than the second-place U.S. emissions. India is now third in emissions. Since 1990, the emissions of the developed countries have grown only marginally overall, and some countries have even managed reductions. Over that period, U.S. emissions have increased by about 10 or 0.5 percent annually. Over that same period, world CO_2 emissions have risen over 50 or 2.5 percent annually.

There would certainly be less debate about restricting CO_2 emissions if there were not economic costs. But there is a tremendous debate how costly restrictions on CO_2 emissions will be. Some argue that developing alternative energy and other activities will offset the cost of restrictions. Others argue that higher energy prices and other costs will cause serious economic damage. Whatever the correct answer, the global economic downturn since 2008 and the weak recovery have made most countries unwilling to risk their economies.

President Obama tells an audience of ardent environmentalists that global warming is real, caused by human activities, and a growing danger. Now is the time to act, he urges. Senator James Inhofe disagrees in the NO selection. He argues that the science regarding the causes and effects of global warming is not clear and that the policies the president advocates to curb U.S. emissions are [redundant] destructive to the U.S. economy. He opposes the move to limit coal as an energy source. Those who agree with the senator point out, for example, the use of abundant and cheap, but CO_2 emitting, coal to produce electricity helps keep its average price per kilowatt hour in the United States at $0.12 compared, say, to $0.35 in Germany and $0.26 in Japan.

YES ↵

Barack H. Obama

Remarks at the League of Conservation Voters Capital Dinner

I know you think I'm here [at the League of Conservation Voters (LCV) dinner] just because I care about the environment. No, it's deeper than that. When I ran for the U.S. Senate [in Illinois], I was decidedly the underdog, really nobody knew me. And LCV, because it's a good-government type goes through process and they had the board interview all the candidates. And I went in and I did my shtick, and they endorsed me. And I was not at all favored to win, and it was the first and probably only prominent national organization to endorse me in the primary; everybody endorsed me in the general. But for me, at least, it was a testament that this was an organization that cared about ideas, and obviously had a really good eye for talent. So I am here primarily out of loyalty. There's a little payback going on here. But then there is also the whole protecting the planet thing.

The work you do to protect our planet and our country, and dealing with the rapidly growing threat of climate change is even more urgent and more important than the last time I spoke to you back in 2006 when I was still a senator. Because we know two big things: We know more about the threat than we did back then, and we know through experience that we can act; that we don't have to be passive, that we can act in ways that protect our environment and promote economic growth at the same time. We know we can do it. We've shown we can do it.

So exactly one year ago today, I was at Georgetown University to announce my Climate Action Plan. And I remember this because it was 95 degrees. The staff purposely put the speech outside and so there are a number of photographs of me wiping my brow, and I don't sweat usually. I was hot. But I started my speech the same way I start all my speeches on climate change—with the facts. Not a lot of spin, just the facts.

We know that burning fossil fuels releases carbon dioxide. We know that carbon dioxide traps heat. We know that the levels of carbon dioxide are higher than they've been in 800,000 years. We know that the 20 warmest years on record for our planet all happened since 1990—and last month was the warmest May ever recorded. We know that communities across the country are struggling with longer wildfire seasons, more severe droughts, heavier rainfall, more frequent flooding. That's why, last month, hundreds of experts declared that climate change is no longer a distant threat—it "has moved firmly into the present." Those are the facts. You can ignore the facts; you can't deny the facts.

So the question is not whether we need to act. The overwhelming judgment of science, accumulated and measured and reviewed and sliced and diced over decades, has put that to rest. The question is whether we have the will to act before it's too late. Because if we fail to protect the world we leave our children, then we fail in the most fundamental purpose of us being here in the first place.

For more than 40 years, that has been your mission: preserve and protect this planet we call home. And by the way, it's been the mission of a lot of members of Congress who are here today. It's been a priority of mine for as long as I've been in office. And part of it maybe is growing up in Hawaii, where every day you appreciate the wonder of your planet but you also understand how fragile it is. So we're working in a few ways to do our part—by using more clean energy, less dirty energy, and wasting less energy throughout our economy.

Right now, America generates more clean energy than ever before. Thanks in part to the investments we made in the Recovery Act. Remember that old Recovery Act? It was the largest investment in green energy and technology in U.S. history—that was just one of its attributes.

As a consequence of those investments, the electricity we generate from wind has tripled since 2008. The energy we generate from the sun has increased more than tenfold. Every four minutes, another American home or business goes solar. And last year alone, solar jobs jumped 20 percent.

And the good news is we can do even better. So my Climate Action Plan will help us double our electricity from renewable energy again by 2020. And I directed the Interior Department to green-light enough private

Barack H. Obama. Speech at League of Conservation Voters Capital Dinner, June 25, 2014.

renewable energy capacity on public lands to power more than 6 million homes. The Department of Defense—the biggest energy consumer in America—is installing 3 gigawatts of renewable power on its bases. So we are going to continue to incentivize the adaptation of technologies that are not going to solve our entire problem, there's no silver bullet, but what we're seeing is unit costs go down, efficiency and power generation going up. We're moving—and it's making a difference.

So that's the first part of our plan: generating and using more clean energy. Then we've got to use less dirty energy. Since I took office, we've doubled how far our cars and trucks will go on a gallon of gas by the middle of the next decade. We're helping families and businesses save billions of dollars with more efficient homes, and buildings and appliances. By the end of the next decade, these combined efficiency standards for appliances and federal buildings will reduce carbon pollution by at least 3 billion tons compared to when I took office, and that's an amount equal to what our entire energy sector emits in nearly half a year.

So together, we've held our carbon emissions to levels not seen in about 20 years. And since 2006, no country on Earth has reduced its total carbon pollution by as much as the United States of America.

And by the way, the private sector knows how important this is. Today, at the White House, some of America's leading foundations and impact investors committed more than $300 million to accelerate clean energy technology and energy-efficient buildings. So we're making progress on that front.

But everybody here knows, for the sake of our kids, we have got to do more. Today, about 40 percent of America's carbon pollution comes from our power plants. There are no federal limits to the amount those plants can pump into the air. None. We limit the amount of toxic chemicals like mercury, and sulfur, and arsenic in our air and water, but power plants can dump as much carbon pollution into our atmosphere as they want. It's not smart, it's not right, it's not safe, and I determined that it needs to stop.

So that's why, in my speech a year ago, I directed the EPA [Environmental Protection Agency] to build on the efforts of a lot of states, and cities and companies, and I told them, come up with commonsense standards for reducing dangerous carbon pollution from our power plants. Last month, I unveiled those proposed standards, which will cut down our carbon pollution, and our smog, and our soot that threaten the health of our most vulnerable Americans, including children and the elderly. We've constructed it so that states have the flexibility to meet

these standards with whatever clean energy sources make sense for them, including renewables and taking advantage of natural glass—natural gas, replacing even dirtier energy sources. And in just the first year that these standards go into effect, they'll help avoid up to 100,000 asthma attacks, about 2,100 heart attacks—those numbers keep on going up after the first year. And we're taking a whole bunch of carbon out of the atmosphere.

So I say all this to say that, no matter how big a problem, progress is possible. It's not instantaneous; we've got to sometimes cut these things into pieces.

It's pretty rare that you encounter people who say that the problem of carbon pollution is not a problem. You've all—in most communities and work places, et cetera, when you talk to folks, they may not know how big a problem, they may not know exactly how it works, they may doubt that we can do something about it, but generally they don't just say, no, I don't believe anything scientists say. Except where?

[The audience responded]: Congress!

[The president continued]: In Congress. Folks will tell you climate change is a hoax or a fad or a plot. It's a liberal plot. And then most recently, because many who say that actually know better and they're just embarrassed, they duck the question. They say, hey, I'm not a scientist, which really translates into, I accept that manmade climate change is real, but if I say so out loud, I will be run out of town by a bunch of fringe elements that thinks climate science is a liberal plot so I'm going to just pretend like, I don't know, I can't read.

I mean, I'm not a scientist either, but I've got this guy, John Holdren [Director of the White House Office of Science and Technology Policy], he's a scientist. I've got a bunch of scientists at NASA [National Aeronautics and Space Administration] and I've got a bunch of scientists at EPA. I'm not a doctor either, but if a bunch of doctors tell me that tobacco can cause lung cancer, then I'll say, okay. Right? I mean, it's not that hard.

Now, the good news is, the American people are wiser than this. Seven in ten Americans say global warming is a serious problem. Seven in ten say the federal government should limit pollution from our power plants. And of all the issues in a recent poll asking Americans where they think we can make a difference, protecting the environment came out on top. We actually believe we can do this. We can make a difference.

And that's in large part thanks to you. Many of you have done just terrific work at the grassroots level—educating, mobilizing. That isn't to say, by the way, and I say this sometimes to environmental groups, that's not to say that it's not easy and that we should not take seriously the very real concerns people have about their current economic state. People don't like gas prices going up. They don't like electricity prices going up. And we ignore those very real and legitimate concerns at our peril, so if we're blithe about saying this is the defining issue of our time but we don't address people's legitimate economic concerns then even if they are concerned about climate change, they may not support efforts to do something about it. So we've got to shape our strategies to speak to the very real and legitimate concerns of working families all across America. But we can do that, that's the good news, we can do it.

And the sooner we do it, the better. Right now, developing countries have some of the fastest-rising levels of carbon pollution. They are less equipped to cope with the effects of climate change than we are. But they're also trying to deal with hundreds of millions of people in poverty. And so the tradeoffs for them are even tougher than for us sometimes unless we describe how development should leapfrog some of the old technologies, learn lessons from us, and go right to a clean energy future. And we should be part of that conversation, but we've got to lead by example. They're waiting to see what America does. And I'm convinced when America proves what's possible, other countries are going to come along.

I don't have to tell you all this. You understand our mission. You've helped define it. And it's not going to happen overnight. This is a generational project. And sometimes it can be easy to get discouraged, and to feel like, oh, we're not setting high enough goals, we're not reaching them quickly enough—I know. I read the science. I'm not a scientist, but I read it. But what I also know is, is that when you take those first steps, even if they're hard and even if they're halting sometimes, that you start building momentum and you start mobilizing larger and larger communities. And when it comes to a challenge as far-reaching and important as protecting our planet, every step makes a difference.

And one of the great things about it is that this is a generational fight but the younger generation is more attuned to this than just about anybody. You talk to Malia, you talk to Sasha [the president's daughters], you talk to your kids or your grandkids, and this is something they get. They don't need a lot of persuading. They understand how important this is. And that should make us hopeful and optimistic.

And I'll close with a story I heard recently that illustrates the point. I called Gregg Popovich, coach of the San Antonio Spurs to congratulate him on winning the NBA [National Basketball Association] Championship. For more than a decade, Coach Pop has hung a sign in the Spurs locker room for all his players to see. And on that sign is a quote from a 19th-century reformer, which is not what you'd expect to see in an NBA locker room but that's the kind of guy Coach Popovich is, and the quote goes something like this: "When nothing seems to help, I go look at a stonecutter hammering away at his rock perhaps a hundred times without as much as a crack showing in it. Yet at the hundred and first blow, it will split in two, and I know it was not that blow that did it, but all that had gone before." [The quote is attributed to Jacob Riis, American journalist and social reformer, 1849–1914.]

So that's what we're doing—together, we are pounding the rock. And together, we are making progress. And sometimes it feels like, man, I'm getting tired. And we're not moving fast enough. But then one day, the rock splits open—not because one person comes up or one President comes up and strikes a mighty blow, but because of all the work that has gone on before. Our work. So until the day comes that the rock is split, we've all got to take turns pounding. We've got to keep fighting. We've got to keep mobilizing. We've got to keep making sure that your voices are heard in Congress, in state capitals, in city halls. Because that's the only way we're going to build the kind of future that we want—cleaner, more prosperous, more good jobs; a future where we can look our kids in the eye and tell them we did our part, we served you well, we were good stewards, we're passing this on.

BARACK OBAMA is the 44th president of the United States and former U.S. senator (2005–2008). He has a JD degree from Harvard University.

James Inhofe

A Speech on the Floor of the U.S. Senate

It is a little bit humorous to me that we are talking about extending unemployment benefits in the midst of one of the most intense cold fronts in American history. I saw one newscaster yesterday who said: If you are under 40, you have not seen this stuff before. It has to make everyone question—and I am going to tie this together—whether global warming was ever real.

While I know the leftwing media is giving me a hard time for talking about my opposition to the administration's global warming policies when it gets cold outside, I think it is important to point out two things. No. 1, the administration is intentionally ignoring the most recent science around global warming, and No. 2, global warming policies costing between $300 billion and $400 billion a year, along with the rest of the EPA's environmental regulations, are resulting in millions of job losses.

We are talking about extending unemployment benefits, yet it is really jobs we need, and the jobs are being robbed from us by the overregulation that is taking place in the Environmental Protection Agency, and of course, the crown jewel of all of those is cap and trade. When I say $300 billion to $400 billion a year, that would constitute the largest tax increase in American history.

I find that sometimes when we are talking about these large numbers it is hard to relate that to everyday people, to our own states, and to how it affects our families. So at the end of each year I get the total number of families in my State of Oklahoma who filed a Federal tax return and I do the math. In this case, it would cost about $3,000 for each family in my State of Oklahoma to pay this tax, this cap-and-trade tax that supposedly will stop us from having global warming.

It is interesting that people now realize this would not stop it. Even if we did something in the United States, it wouldn't affect overall emissions of CO_2 [carbon dioxide]. And that is what we are talking about. That is what makes global warming so important to mention as we debate the extension of unemployment benefits.

If we want to improve our employment figures, what we need to do is stop the onslaught of environmental regulations that have come out during this Obama presidency.

First, let's talk about the global warming issue. It is interesting that we have often seen global warming related to events affected by unseasonable or unusually cold weather. Often, this has occurred whenever [former vice president and environmental protection advocate] Al Gore has been involved in an event. Let me give a couple of examples. In January of 2004, Al Gore held a global warming rally in New York City. It turned out to be what would go down as one of the coldest days in the history of New York City. Three years later, in October of 2007, Al Gore gave a big global warming speech at Harvard University, and it coincided with temperatures that nearly broke Boston's 125-year-old temperature record.

In March of 2009, Speaker of the House Nancy Pelosi was snowed out of a global warming rally in Washington, DC. Because of all the snow, her plane wasn't able to land and they had to cancel her appearance at the event. A year later, in March of 2010, the Senate Environment and Public Works Committee had to cancel a hearing entitled "Global Warming Impacts in the United States" due to a major snowstorm. At that time, I was the ranking member of that committee, and they were all geared up, ready to have this big hearing, and they couldn't do it because of a major snowstorm. That was in 2010. So this has been going on now every year going back to 2010.

Just last year, in July of 2013, a cruise liner that was chartered to discuss the impact of global warming planned to sail through the Northwest Passage of the Arctic but got stuck because the passage was full of ice. Now, more on that in a minute. In that same month, Al Gore had an event in Chicago training people about global warming but was greeted with the coldest temperatures in 30 years.

A lot of folks, even in the last day, have said that just because there are cold temperatures does not mean global warming has stopped. Most alarmists will, however, correct you that it's no longer global warming, but instead,

James Inhofe. Speech to U.S. Senate, January 8, 2014.

climate change. Increases in temperature still matter. In a November 2013 executive order, the President implemented new climate change policies—very expensive ones; large tax increases—stating that "excessively high temperatures" are "already" harming natural resources, economies, and public health nationwide. In other words, he's implementing his climate change policies because of rising temperatures, otherwise known as global warming.

So temperatures falling and really cold days do matter. It does matter when the ice caps are growing and temperature increases pause for 15 years. And that is what has happened for the last 15 years. If global warming is not happening, then there is no need for the ensuing policies—whether you call it global warming or anything else.

Monday was a cold day. At one point, the temperature average in the country was 12.8 degrees. In Chicago, it was 16 degrees below zero. That broke the record that was set way back in 1884, when it was 14 degrees below zero. This made Chicago colder than even the South Pole at the same moment, where it was only 11 degrees below zero.

Just this week, down at the South Pole, a number of ships were stuck in the ice, even though it is in the middle of the summer down there. This was all over the news, and for good reason.

On November 27, a research expedition to gauge the effect of global warming on Antarctica began. On December 24, a Russian ship carrying climate scientists, journalists, tourists, and crew members for the expedition became trapped in deep ice up to 10 feet thick. An Australian icebreaker was sent to rescue the ship, but on December 30 efforts were suspended due to bad weather. On January 2, a Chinese icebreaker, the *Xue Long*, sent out a helicopter that airlifted 52 passengers from the Russian ship to safety on the Australian icebreaker. The Chinese vessel is now also stuck in the ice along with the Russian vessel. Twenty-two Russian crew members are still on board the Russian ship, and an unreported number of crew members remain on the Chinese ship. On January 5, the Coast Guard—that is us; we came to the rescue—called to assist the ships that are stuck in the Antarctic. Our icebreaker ship is called the *Polar Star*.

Just a few months ago the journal *Nature*—that is a well-respected publication on environmental science—they published an article that said over the last 15 years "the observed [temperature] trend is . . . not significantly different from zero [and] suggests a temporary 'hiatus' in global warming." This is not something that is appreciated by the Obama administration. What they are saying is—and this was the journal—that it had stopped. In fact, I along with some of my colleagues have asked the President

for the data backing up his claims that warming is actually happening faster now than previously expected. Considering the most recent data, those statements have not been true. No models predicted there would be a fifteen-year pause in global warming, but the President hasn't yet fully responded to our inquiry. Let's go back. When you look back in history, and you look at these cycles, you have to come to the conclusion that God is still up there.

I have this from memory, and I think I will get this right. From 1895—they had a cold spell that came in, and that is when they said another ice age is coming. That lasted until 1918. In 1918, that all changed, and all of a sudden it started getting warmer, and that is when the term "global warming" first came out. So from 1918 to 1945 it was a warming period that we went through. Then, in 1945, it changed and another ice age was coming that everyone was concerned about. That lasted from 1945 to 1975.

Then, in 1975—and this is interesting because in 1975 we got into this time period we are talking about now; and that is, they were saying that global warming is coming upon us. Well, what is happening now—and these people have an awful lot of their time and resources and reputation at stake here—it is now to the point where that has reversed and we are going into another one of these cycles.

The interesting thing about 1945 is that 1945 was the year where the greatest surge in CO_2 emissions happened. It was during that year. That was right after World War II. That precipitated not a warming period but a cooling period.

In December of 2008, Al Gore said, "The entire North Polar ice cap will disappear in five years." The North Polar cap is the Arctic ice cap.

Well, we are now 5 years later when, as Al Gore said, it should all be melted by now. The deadline was December of 2013. Arctic ice is actually doing pretty well. Just last month, the BBC [British Broadcasting Corporation] reported that the Arctic ice cap coverage is "close to 50% more than in the corresponding period in 2012." In other words, in 1 year it increased by 50 percent. This is the very ice cap that Al Gore said would be gone by now. So contrary to what Al Gore predicted, the ice cap did not disappear last year; it grew.

In May of 2006, Al Gore said in his movie, *An Inconvenient Truth*, that the Antarctic ice cap melt could result in a 20-foot increase in sea levels. You contrast that with the frozen global warming expedition down there this week and a September 2013 report in the *Washington Post* that Antarctic sea ice has hit a 35-year high this past year.

Now, these things—people do not seem to stop and think. These were predictions that were made. This is the same Al Gore where there was an article in the *New York Times* saying that arguably he is the world's first environmental billionaire, and all these things people were saying were gospel truth. Now we know they are not, but nobody talks about it. The media does not talk about it. When you put it all together, it is impossible not to sit back and wonder: If there is this evidence that the temperatures are actually getting colder, should we really pursue cap and trade and other similar regulations and policies that will cost the economy $300 billion to $400 billion a year to implement? In light of our high unemployment levels—and that is what we are talking about today; we are talking about extending unemployment insurance—I do not think so. That is what we are here talking about anyway: unemployment numbers.

To help remedy the problem, I am submitting two amendments. The first one I want to talk about is amendment No. 2615.

The EPA has systematically distorted the true impact of its regulations on job creation by using incomplete analyses to assess the effects of its rules on employment. They have even published that many of their regulations will result in net job creation.

EPA's costly regulations, as any reasonable person knows, actually reduce business profitability and cause actual job losses. New mandates and requirements do not help the economy add jobs. For example, the EPA estimated that its 2011 Utility MACT—that was passed. MACT means "maximum achievable control technology." In other words, we come along in all of our great wisdom up here and we pass a law saying how much emissions can take place, and yet there is no technology that will accommodate that.

So the EPA estimated that its 2011 Utility MACT—that is the one that passed; it was passed into law—rule would create 46,000 temporary construction jobs and 8,000 net new permanent jobs. By contrast, a private study conducted by NERA Economic Consulting that examined the "whole-economy" impact of the rule—and we are talking about the Utility MACT; that is what put coal out of business in a lot of the United States—the study estimated that the rule would have a negative impact on worker incomes equivalent to 180,000 to 215,000 lost jobs in 2015, and the negative worker impact would persist at the level of 50,000 to 85,000 such "job-equivalents" annually.

The EPA estimated its Cross State Air Pollution rule would create 700 jobs a year. By contrast, the same NERA study estimated the rule would eliminate 34,000 jobs from 2013 through 2037. It lets you know that the EPA is controlled by the President, and they are there to fortify

anything he says, even though we have studies to show just the opposite is true.

The EPA also estimated its Industrial Boiler MACT rule—every manufacturer has a boiler, so this affects all manufacturers—would create 2,200 jobs a year. By contrast, NERA, in their study, estimated the rule would eliminate 28,000 jobs each year from 2013 to 2037.

In addition to those examples, the National Association of Manufacturers did a study that determined the cumulative impact of EPA's regulations is $630 billion annually and totals about 9 million jobs lost. That did not even include the cap-and-trade regulations, which would cost another $300 billion to $400 billion per year.

The EPA has not yet fully studied or disclosed the impact of these rules, but we know it is going to be very expensive.

If we really want to do something about unemployment numbers in this Nation, we need to hit the brakes on EPA's regulations. Let's do not worry about extending the time of unemployment compensation, unemployment insurance; let's do something about the costly regulations.

So my amendment does this by prohibiting the EPA from making any of its new regulations final until it complies with requirements under the Clean Air Act's section 321.

Section 321 was put into the Clean Air Act back in 1977, and it was supposed to require the Federal Government to state what the job impact would be as a result of the various regulations it pursued. How many times has the EPA conducted this study? Not once. So that amendment would help reduce the impact of EPA's rules on job loss.

My second amendment would actually help create jobs. It is really kind of unrelated, but since I am talking about two amendments that are very significant now and would help resolve our jobs problem to a great extent, I will talk about amendment No. 2605. It would help us take advantage of our vast domestic oil and gas resources.

We have seen huge increases in oil and gas development in recent years due to the advancements in precision drilling, hydraulic fracturing, and other technologies. These technologies have unlocked the shale revolution and, because of this, official government estimates now predict that we will become completely energy sufficient by 2035.

What they will not tell you is that this could happen a lot faster. Right now, 83 percent of Federal lands are currently off limits to oil and gas developers. There is not a good reason for this. It is just the administration preventing us from having more jobs and energy independence.

You have to keep in mind, we now and then hear people from the Obama administration saying: Well, wait a minute, during the last 4 years or 5 years, the production has increased by some 40 percent. But that is all on State property and on private land. On Federal land, it has actually decreased by about 15 percent because of the war against fossil fuels that has taken place out of the White House.

So the amendment I am offering would give these resources to the States to unlock and develop on their own. The assumption here is the States should be in a better position to know what they want to do with these regulations in their own State and any damage that might come to the environment—let them make that decision instead of the Federal Government doing it.

A recent report by the Institute for Energy Research estimated that if we completely developed these off-limits Federal resources, it would create 2 ½ million jobs and generate $14.4 trillion in economic activity. But it would also help us achieve energy independence by 2024, 11 years sooner than it would otherwise.

So if we want to create jobs, this is how we can do it. We should embrace our energy future and aggressively expand production. If we want fewer people to lose their jobs in the future, we should prevent the EPA's regulations from moving forward, at least until they fully study the impact the rules will have on job losses.

We have been trying to do this now for a long period of time, to determine what these costs are. When the American people find out, in terms of the dollars of cost and the jobs that are lost with excessive regulation, they will come and let their feeling be known, certainly at election time.

JAMES INHOFE is a U.S. senator from Oklahoma, and member of its Committee on Environment and Public Works. He received a BA degree from the University of Tulsa.

EXPLORING THE ISSUE

Is President Obama's U.S. Global Warming Policy Wise?

Critical Thinking and Reflection

1. What has been your perceived experience of global warming?
2. There has been a tendency among Americans, at least, to support restraints on CO_2 emissions and other steps to curb global warming but to resist paying the lifestyle or economic price to implement many of the steps such as higher gasoline taxes to suppress use. How far would you be willing to go to support such steps?
3. Nuclear energy provides one available solution to CO_2 emissions. How would you feel about a massive effort to replace coal-fired plants that produce electricity with nuclear plants?
4. Should the United States make a substantial effort to reduce carbon dioxide and other greenhouse emissions unless China, India, and other newly industrializing and modernizing countries also do so?

Is There Common Ground?

Finding common ground will be difficult. Certainly it is; electricity driven by solar, wind, thermal, and other green sources is expanding and becoming cheaper than it once was. But those sources combined have remained more expensive than coal-fired plants and have other drawbacks. Nuclear generating plants are emissions free, but they are expensive and present some safety hazards, although how much is subject to debate. It may be that jobs created by "going green" will offset those lost to having higher energy costs and therefore higher product costs than competing countries, but that remains an optimistic projection at best. It is also true that the poor and those on fixed incomes (the elderly) are often the ones to feel the costs of increased energy most directly and severely. Nevertheless, if the president and most of the scientific community are right, then global warming will have an increasingly direct impact on our lives and finances.

It is also true that nothing the United States can realistically do will have a significant impact on global warming, even if humans are entirely responsible for it.

Emissions from China, India, and elsewhere are the most important factor today, and the best that even strong efforts by the United States can do is to slow down the global growth of emissions. That would delay, but not change, the dire consequences from global warming that some predict.

Create Central

www.mhhe.com/createcentral

Additional Resources

Victor, David G. (2011). *Global Warming Gridlock: Creating More Effective Strategies for Protecting the Planet* (Cambridge University Press)

Giddens, Anthony. (2011). *The Politics of Climate Change* (Polity Press)

Wolinsky-Nahmias, Yael. (2014). *Changing Climate Politics: U.S. Policies and Civic Action* (CQ Press)

Internet References . . .

Climate Change—The White House

www.whitehouse.gov/energy/climate-change

Environmental Defense Fund—Fight Global Warming

http://www.fightglobalwarming.com

Environmental Protection Agency—Climate Change

http://epa.gov/climatechange/ index.html.

Global Warming—May Cooler Heads Prevail

http://www.globalwarming.org/

United Nations—Global Warming

http://www.un.org/wcm/content/site /climatechange/gateway

Selected, Edited, and with Issue Framing Material by:
John T. Rourke, *University of Connecticut, Storrs*

ISSUE

Should the United States Deport Unauthorized and Unaccompanied Immigrant Children?

YES: Dan Coats, from "A Speech on the Floor of the U.S. Senate," *Congressional Record* (2014)

NO: Mark Seitz, from "Crisis on the Texas Border: Surge of Unaccompanied Minors," Testimony During Hearings before the Committee on Homeland Security, U.S. House of Representative (2014)

Learning Outcomes
After reading this issue, you will be able to: • Understand the opposing approaches to and priorities regarding immigration. • Differentiate the issues in the debate over unauthorized immigrants in general from the debate over unaccompanied children. • Explain how Bishop Seitz's recommendations about how the United States can begin to remedy the problem of unaccompanied children extend beyond the immediate status and treatment of the children to include U.S. policies to ease the violence and poverty that he claims are the cause of the children's flight from their home countries.

ISSUE SUMMARY

YES: Dan Coats, a U.S. senator from Indiana, argues that allowing the flood of unaccompanied children who have arrived illegally in the United States will only encourage the dangerous practice of sending them to the United States.

NO: Mark Seitz, the Roman Catholic bishop of the Diocese of El Paso, Texas, and head of the Committee on Migration of the United States Conference of Catholic Bishops, tells a committee of Congress that a faithful adherence to the "best interest of the child" standard should govern how all undocumented immigrant children are treated. In most cases Seitz argues the best interest of the child is to remain in the United States to avoid returning them to lives of poverty and, often, violence.

Immigration has become a hot topic in the United States in recent years. There has been a significant increase in the number of immigrants coming to the United States and the percentage of residents of the United States who are immigrants.

Legal immigration nearly tripled from an annual average of 330,000 in the 1960s, to an annual average of 978,000 in the 1990s, to an annual average of just over a million from 2000 through and 2013. These recent numbers seem huge, but they are approximately the same as the yearly number of legal immigrants between 1905 and 1914, a period when the U.S. population averaged about 90 million, less than 30 percent of the average U.S. population in recent years. Moreover, during the earlier period, legal immigrants equaled about 1.1 percent of the population, whereas the annual number of legal immigrants has equaled only about 0.3 percent of the American population during recent (2000–2013) years.

This data does not tell the complete story of immigration, though. Adding to the number of individuals entering the country is the sizable number of immigrants living in the United States who have come without legal permission from the U.S. government. These immigrants are called "illegal immigrants" by some, especially those who want to reduce immigration and/or deport many of them, and called "undocumented immigrants" or "unauthorized immigrants" by those who take a more benign view of them and believe that most should be able to become citizens. It should be noted that children born in the United States are citizens, no matter what the status of their parents is (including mere visitors). Thus many immigrant families include parents without their immigration papers and children who are fully American citizens.

Estimates of the illegal/unauthorized population vary quite a bit, but numbered according the estimate of U.S. Department of Homeland Security (DHS, including the three main agencies that deal with immigration: Citizenship and Immigration Services, Immigration and Customs Enforcement, and the Border Patrol), there were 11.4 million undocumented immigrants living in the United States in 2012. As such, these immigrants constitute about 27 percent of the foreign-born residents of the United States and about 3.5 percent of the entire U.S. resident population. Among legal immigrants, about 42 percent have become citizens, with the remainder eligible for citizenship. Of course, none of the illegal/unauthorized immigrants are eligible for citizenship.

Although it is seldom voiced openly, the origin of immigrants is another factor that has added to the immigration debate. As recently as the 1950s, more than 70 percent of all documented immigrants coming to the United States originated in Europe, Canada, and other predominantly "white" countries. Changes in immigration law allowed increasing numbers of immigrants from countries in Latin American, the Caribbean, and Asia, with countries whose populations are mostly made up of "people of color." Now, about 80 percent of all legal immigrants come from these "nonwhite" countries, as do virtually all illegal immigrants. Among this group, DHS estimates that 59 percent are from Mexico and another 15 percent from Central America.

There have almost certainly been unauthorized immigrants in the United States since Congress first regulated immigration and citizenship in the Naturalization Act of 1790. But the number of these newcomers became an increasing source of debate beginning in the mid-1960s. Up to then, and until the United States ended the Bracero program, several hundred thousand Mexicans were allowed entry on temporary work visas. However, Congress ended that program in 1964 amid charges from one point of view that Bracero workers were being abused and charges from another point of view that they were taking jobs away from Americans.

Irrespective of the new law, Mexicans seeking economic opportunities and Americans business seeking cheap labor combined to not only maintain, but increase the flow of Mexican and also Central American illegal/unauthorized immigrants. By 1980, the U.S. Census Bureau estimated that between two and four million such immigrants, mostly Latinos, were in the country. Congress sought to deal with this group by passing a law in 1986 that allowed unauthorized immigrants who had entered the United States before 1982, who had been continuous residents, and who had met other requirements (such as no conviction of a felony) could apply for legal residency and eventually citizenship. Supporters claimed that this helped alleviate the problems such people face. Critics charge the amnesty merely encouraged more illegal/unauthorized immigration. The terror attacks of 9/11 and subsequent concern with terrorism has also come into play, with some arguing that potential terrorists could enter the United States unless stronger border security is put into place.

Whatever anyone's views, still more illegal/unauthorized immigrants have entered the country, and the debate has escalated. There have been some laws and executive actions easing the conditions of such immigrants and other laws strengthening border security to keep them out and increasing effort to deport them. But Congress and the president have not been able to agree on comprehensive reform. In recent years, the debate over immigration has become further confounded by an unprecedented increase in the number of unaccompanied and undocumented immigrant children arriving in the United States, mostly across the U.S./Mexico border. These children come from all over the world but predominately from Guatemala, El Salvador, Honduras, and Mexico. Whereas in fiscal years (FY) 2004–2011, the number of unaccompanied children apprehended by the U.S. government averaged around 6,000–8,000 per year, the total jumped to over 13,000 in fiscal year 2012 (FY, October 2011–September 2012), over 24,000 in FY 2013, and an estimated 60,000 in FY 2014.

The following selections touch on a wide range of issues such as whether to treat these children as refugees or illegal immigrants, how to house and otherwise care for them, and how to adjudicate their individual cases. The bottom line debate is whether to allow some or all of the children to remain in the United States. Senator Dan Coates says that the most common answer should be to return the children to their home country and parents or other guardians. Bishop Mark Seitz urges American to respond with compassion "to the needs of these children . . . [and] not turn them away or ostracize them."

YES ⬅

Dan Coats

A Speech on the Floor of the U.S. Senate

As have many Americans, I have watched with increasing concern and increasing frustration the rapidly growing humanitarian crisis on our southern border. More than 60,000 unaccompanied alien children—mostly minors from Guatemala, Honduras, and El Salvador—have been apprehended at the border in this fiscal year, and we have 2½ months remaining. The numbers are staggering. Another 40,000 family members—one or both parents traveling with their children—have also been apprehended just in this fiscal year. [The fiscal year, the U.S. budget year, extends from October 1 of one calendar year to September 31 of the next. The date of the fiscal year, such as fiscal year (FY) 2014 corresponds to the end-date of that year.]

To put these numbers in perspective, in 2008, the number of unaccompanied alien children apprehended at the border was 8,000. Three years later, in 2011, the number had doubled. It had doubled to 16,000. This is a situation we perhaps didn't see coming, but should have.

Today, of course, the numbers are staggering, as I mentioned. The number has skyrocketed. In fact, in April and May of this year, 10,000 have arrived. We simply cannot sit back and let this situation grow worse as it does day by day. We must now find a way to solve this crisis and stem the flow of unaccompanied minors entering our country. It is imperative that this Congress and this administration work together to do this and do this immediately. We dare not move toward our regularly scheduled August recess without accomplishing the solution or resolution of this current crisis, which is impacting children, impacting families, impacting communities, impacting many across the United States in terms of this crisis.

As we do this, I think it is important that we be guided by some key principles, including laws that are currently on the books—laws that might need to be adjusted—as well as compassionate hearts in terms of how we deal with those who are here but will need to be returned to their homeland.

First, clearly and foremost, we have to enforce existing law. Existing law says we need an orderly process.

Immigration needs to be legal. It needs to be processed in an orderly way and in a way so that we can accommodate those who come from out of the country. I am the son of an immigrant who was processed through a legal process, a process that speaks for many of us not only here in this chamber [the Senate] but for many across America. We are all in a sense immigrants. For over 200 years, we have come as immigrants through a legal process. Today we find a situation where our borders are being swamped with those who are attempting to come illegally, for whatever reason. More importantly, we have to make it clear to them that the law does not allow this to happen. So we have to get control of the border. We have to get control of our immigration process.

I think all of us feel the need for immigration reform. Step No. 1 has to be securing our borders so we can convince the American people we can return to an orderly process of bringing immigrants to this country and not be overwhelmed by the illegal immigration flowing to our southern borders. It is also important because we need to let the families know and the children know their trip to America is not what has been promised them.

Many believe this humanitarian crisis is focused on how we handle these children once they arrive at the border, and there is a need to address that issue. But in reality, the crisis for these children begins when they start their trip, given the dangers of the journey. We now know the children who are making these dangerous treks from Central America are often in the hands of smugglers, drug cartels, coyotes—criminal elements that are delivering a false lie to families and individuals in these countries. They are basically saying, Get your children across the border and they will then be absorbed into American society and they will be in a better place. And, by the way, write us a check for $7,000 or $10,000 or $5,000, whatever the market bears, and we will ensure that your children arrive safely, and then you won't have to worry about them anymore. That is simply not true.

Sadly, from the latest information that has come to us, in surveys that are being taken and investigations that

Dan Coats. Speech to U.S. Senate, July 15, 2014.

are being made, the story is horrendous. Often, for those in the hands of those who are seeking to bring them along the approximately 1,500-mile trip from Central America to the Texas border, the reality of what these children are facing and what these families are facing is startling and it is an issue that absolutely has to be addressed.

Doctors Without Borders exists in southern and central Mexico, and they did surveys of those who were attempting to make this trip. They indicated that 58 percent of their patients suffered at least—at least—one episode of violence along their way from Central America to the United States. One media network did an investigation that followed the path of Central American migrants, including children, and while their numbers have not been verified or documented, they are staggering. Even if the results are half of what they claim, it is a situation of immense humanitarian dysfunction. They found that 80 percent of all migrants will be assaulted, 60 percent of women will be raped, and only 40 percent will actually make it to the border.

Let's say those numbers are exaggerated. There is some indication this media outlet was, perhaps, sensationalizing their numbers. Let's say it is just half of that. But if it is half of that, it is a situation we absolutely cannot tolerate. We absolutely cannot sit by and say the only humanitarian crisis is taking care of these children once they cross the border—making sure they have vaccinations, sustenance, and a place to sleep until we get them processed. Those who claim that need to understand the crisis that exists before they ever get to the border, and the impact on these children in particular.

In 2010, when the narrative coming out of the administration was chipping away at our nation's immigration laws through the abuse of prosecutorial discretion, this generated whispers of hope that ran rampant through the families of our Central American neighbors and gave a false confidence that if you illegally enter our country, once you are here, you will be able to stay. The belief spread in 2012 when the President [Barack Obama] took his prosecutorial discretion a step further by essentially halting the removal of illegal immigrants who arrived as minors.

There was a process where, of course, they were given a piece of paper, which basically said: You have to appear before a judge, who will determine whether you are able to stay in the country or whether you will have to be sent back home.

The narrative there was: This is your document that allows you to stay in America. In fact, it was not that at all. But because of the overwhelming number of people who received these documents, allowing them to stay here until they were adjudicated by a judge—because that number now exists around 375,000, and there is no way we can possibly adjudicate these and make these decisions in a short amount of time—those who arrived simply melded into the society, and most never showed up before a judge who was making a decision about their legality or illegality.

A key part of what we have to do here, in my opinion, is a repatriation plan. It is easy to just simply throw money out there and say we will come up with a plan later. I cannot support a provision that does not have policy changes to address this situation—policy changes that will allow us to inform our Central American neighbors that they must make every possible effort to engage with us in telling the truth to their constituencies and the parents of these children as to what lies ahead for them: the fact that they will be subjected to potential brutality, unspeakable, brutal efforts and consequences of this trip, as well as returned to their families and their countries.

We have to together make this message clear that our laws require that these children be sent back, but we also have to make it abundantly clear they are putting their children at great harm and great risk to believe this narrative that says: They will be fine, they will be taken care of. Just give us the money and we will make sure your children become Americans and they will be fine in the future.

Secondly, I think we need to go a step further. To deter children from making this journey, we have to return those who have already come.

Included in a viable repatriation program has to be a streamlined process. I mentioned the number of the hundreds of thousands who are still waiting for their adjudication. There have been efforts and suggestions made by some of our colleagues on a bipartisan basis that we address and dramatically increase the number of judges who can go down to the border and make these decisions quickly so we can safely return these children home without having the horror of seeing these children rejected in different communities and no place to put them, as the numbers simply overwhelm our ability to care for them.

The administration does have some flexibility under current law to move families and children through these immigration proceedings in an accelerated manner. However, I believe—and the Secretary of Homeland Security has stated—that we need to go further to change current law to treat all unaccompanied alien children the same.

Now this is the President's own Secretary of Homeland Security, who has been to the border, whom I have met with and talked to several times, who is assiduously trying to address this issue in a bipartisan way. We need to work together to make sure we put the processes in place and the policies in place before we simply decide on a number and hope for the best later.

We need to change the law to allow Central American children who qualify to choose voluntarily to return as well, rather than go through drawn-out immigration proceedings that should still lead to their removal and damage any chance they have to seek legal immigration in the future.

This narrative out there, this story out there, is: Oh well, just go back across the border. Then maybe tomorrow you will get back here, and someone else will pick you up, and you will go to a different place, and you will start the process all over again, and you will finally get handed a piece of paper, and then don't worry about showing up in 12 to 18 months later. You can meld into society, and everything will be well. That absolutely has to be addressed. If we do not do that, we will not succeed with this process.

We also need to use our leverage with these foreign countries to gain their cooperation if they refuse to cooperate with us—whether it is withholding foreign aid, whether it is any number of punitive measures. We need to make sure the governments of these nations understand the risk to their children, the harm to their children, and the fact that we are going to enforce the law, and that if they want to continue future relations with the United States through a legal immigration process, they have to work with us to convince their constituencies and give them the truth as to what is happening to their children—to engage in this process of working with us to stop this flow of illegals.

Now, obviously, we have to provide reasonable care for those who are already here. The vast majority of the new funding the President is requesting would go for caring for the illegal immigrants who are already here. It includes housing, transporting, and caring for the children and families already in the United States.

I believe it is our responsibility as a nation and as a compassionate society to care for the hurt and displaced. But we cannot simply open our arms and encourage all the world's children to strike out on their own, face endless dangers, and come to our shores with the belief that they will be welcomed and accepted and integrated into our society. We simply do not have the capacity to do that on a worldwide basis, and we see the trouble we are having from just three countries. What are we actually doing to stem the flow of unaccompanied alien children coming to the United States? And when will we begin to see the

tide turn? That is something that has to happen and must happen initially.

Finally, in addition to the care which we must provide—the sustenance and the health care and the bedding and the nutrition and the efforts we need to make; and thank goodness for so many nonprofit organizations, churches, and others that have volunteered to join us in this particular effort—but it cannot be an ongoing effort. It has to be something that is accompanied by significant changes I have talked about before in terms of policy. You have to stop the bleeding. You have to stop the effort first and convince the American people that we finally gained control of our borders before we can move to any kind of sensible immigration reform.

This is going to be expensive. We are going to have to make sure the money we are spending is spent as part of a plan to address the problem—not just simply address it and have the problem continue, but address it in a way, on a one-time basis, that we put an end to this story: Send your children and they will be just fine.

The time is moving on, and I know my colleague is waiting to speak and we have votes coming up. So let me shorten this by simply concluding, at the end of the day, we have a huge humanitarian crisis on our hands on our border. I believe we have a moral responsibility to swiftly address and solve this crisis. We have to understand that the crisis involves more than just unaccompanied minors. We cannot ignore the national security implications of a weak border. There are many dark powers in this world that wish to see the influence of the United States diminish—that wish to extinguish the beacon of freedom that we have been to the world.

So for the sake of the rule of law, for the sake of our national security and the safety of these children, it is imperative we act now and get it right. It will only happen if this body, the Congress—the House and the Senate—and the President will work together to put in place, on an expedited basis, a sensible plan to address this humanitarian crisis. "Save the children" means: Don't put those children in the hands of smugglers, coyotes, criminal elements, only for them to go through the horrendous consequences that have become the humanitarian crisis we are addressing.

DAN COATS is a Republican U.S. senator from Indiana and a former U.S. ambassador to Germany (2001–2005). He received a JD degree from Indiana University.

Mark Seitz

 **NO**

Crisis on the Texas Border: Surge of Unaccompanied Minors

I testify today about the humanitarian crisis of unaccompanied child migrants arriving at the US-Mexico Border. I am here to speak with you today about this special population of vulnerable children. In November 2013, I was privileged to lead a United States Conference of Catholic Bishops [USCCB] delegation traveling to Southern Mexico, El Salvador, Guatemala, and Honduras to examine and understand the flight of unaccompanied migrating children and youth from the region and stand in solidarity with these children and their families. During our mission to Central America, we visited migrant children shelters, heard tearful stories from grandmothers waiting to pick up their recently repatriated grandchildren, and listened to children as young as six years old speak solemnly of trafficking and exploitation that was inflicted upon them along their migration journey. The report that came out of our mission acknowledged that a new paradigm regarding unaccompanied children is upon us—namely it is clear that unaccompanied children are facing new and increased dangers and insecurity and are fleeing in response. As a result, this phenomenon requires a regional and holistic solution rooted in humanitarian and child welfare principles. Since our mission and report issuance, many of the humanitarian challenges facing this vulnerable population have persisted and increased. In my remarks, I will highlight and update our observations and recommendations from that report.

My testimony today will recommend that Congress:

- Address the issue of unaccompanied child migration as a humanitarian crisis requiring cooperation from all branches of the US government and appropriate the necessary funding to respond to the crisis in a holistic and child protection-focused manner;
- Adopt policies to ensure that unaccompanied migrant children receive appropriate child welfare

services, legal assistance, and access to immigration protection where appropriate;
- Require that a best interest of the child standard be applied in immigration proceedings governing unaccompanied alien children;
- Examine root causes driving this forced migration situation, such as violence from non-state actors in countries of origin and a lack of citizen security and adequate child protection mechanisms; and
- Seek and support innovative home country and transit country solutions that would enable children to remain and develop safely in their home country.

Overview of the Current Situation of Unaccompanied Children

Since 2011, the United States has seen an unprecedented increase in the number of unaccompanied migrating children arriving at the US/Mexico border. These children come from all over the world but predominately from Guatemala, El Salvador, Honduras and Mexico. Whereas in fiscal years (FY) 2004–2011, the number of unaccompanied children apprehended by the US government averaged around 6,000–8,000 year, the total jumped to over 13,000 in FY 2012 and over 24,000 in FY 2013. ORR initially estimated that about 60,000 unaccompanied minors would enter the United States during FY 2014. Recent government estimates have been revised, projecting 90,000 child arrivals in FY 2014 and 130,000 in FY 2015.

As of June 20, US Customs and Border Patrol (CBP) have apprehended 52,000 in the Southwest Border region for FY 2014. In response to the increased number of unaccompanied children arriving at the US-Mexico border, HHS requested and received approval from the Department of Defense for the use of Lackland Air Force base in San Antonio and a Naval Base in Ventura County in California, which are, respectively, providing shelter to

Mark Seitz. Testimony during hearings on "Crisis on the Texas Border" before the Committee on Homeland Security, U.S. House of Representatives, July 3, 2014.

1,290 and 600 children. Facilities at Fort Sill, Oklahoma, also will house 600 unaccompanied children. The federal government is currently looking at other housing facilities throughout the United States.

With the increasing numbers of unaccompanied children arriving at the US-Mexico border, we must understand who these children are, what is propelling them to travel alone on an increasingly dangerous journey, and what can be done to best address their welfare. Mr. Chairman, I would like to share the stories of three children—one from El Salvador, Guatemala, and Honduras—to give the committee a sense of the reality of the violence they are fleeing:

Factors Pushing Unaccompanied Children to the U.S. Border

In our delegation to Central America in November 2013, USCCB focused upon learning more about the push factors driving this migration and possible humane solutions to the problem. While poverty and the desire to reunify with family to attain security are ongoing motivations to migrate, USCCB found that that an overriding symbiotic trend has played a decisive and forceful role in recent years: generalized violence in the home and at the community and state level. Coupled with a corresponding breakdown of the rule of law, the violence has threatened citizen security and created a culture of fear and hopelessness that has pushed children out of their communities and into forced transit situations.

We acknowledged in our trip report in January that each country exhibited individual challenges which have added to these push factors. Additionally, in response to the increased flow of children in recent weeks, we also acknowledge that certain new country-specific factors may have impacted the latest flow of children. One such factor is the recent crackdown of gang-activity from within prisons in Honduras and efforts to increase police presence by newly elected leader Juan Orlando Hernández. With the increased efforts by the Honduran government to stem communications from gang-leaders within prisons, there are reports of increased violence as gangs fragment and midlevel criminal operators compete for control.

The ongoing generalized violence, leading to coercion and threats to the lives of citizens—particularly children—of these countries, is the overwhelming factor facing these children and propelling their migration. Extortion, family abuse and instability, kidnapping, threats, and coercive and forcible recruitment of children into criminal activity perpetrated by transnational criminal organizations and

gangs have become part of everyday life in all of these countries. In addition to the violence and abuse at the community and national level, transnational criminal organizations, such as the Mexican-based Zeta cartel, which deals in the smuggling and trafficking of humans, drugs, and weapons, operate in these countries and along the migration journey with impunity, and have expanded their influence throughout Central America.

I note that the increase in violence in Guatemala, Honduras and El Salvador forcing children and adults out of their homes is affecting the entire region, not just the United States. For example, since 2008 Mexico, Panama, Nicaragua, Costa Rica, and Belize—the countries surrounding the Northern Triangle countries—have documented a 712% combined increase in the number of asylum applications lodged by people from El Salvador, Honduras, and Guatemala.

In our January trip report we detail the increased violence against children and families in Central America. Given the difficult conditions minors must confront in their home countries, USCCB believes that a robust protection regime for children must be implemented in Central America, Mexico, and the United States. Based on our presence in sending countries, we see the following as reasons for the increased number of children arriving in the United States:

a. Violence perpetrated by organized transnational gangs, loosely-affiliated criminal imitators of gangs, and drug cartels, has permeated all aspects of life in Central America and is one of the primary factors driving the migration of children from the region. USCCB found that in each country—particularly Honduras and El Salvador—organized gangs have established themselves as an alternative, if not primary, authority in parts of the countries, particularly in rural areas and towns and cities outside the capitals. Gangs and local criminal actors operating in Honduras, El Salvador, and Guatemala have consolidated their bases of power, expanded and upgraded their criminal enterprises and honed their recruitment and terror tactics. In many cases, the governments are unable to prevent gang violence and intimidation of the general public, especially youth. USCCB heard accounts of gang members infiltrating schools and forcing children to either join their ranks or risk violent retribution to them or their families. Even in prisons, incarcerated gang members are able to order violence against members of the community. There also were reports that law enforcement have collaborated with the gangs or at least have been lax in enforcing laws and prosecuting crimes. For example, according to Casa Alianza, an NGO that works in Honduras, 93 percent of crimes perpetrated against youth in Honduras go unpunished.

b. Localized violence has severely exacerbated the lack of economic and educational opportunities for youth and has led to stress on the family unit, family breakdown, and even domestic abuse, which leaves children unprotected and extremely vulnerable. The escalation in violence, combined with the lack of jobs and quality education, has led to a breakdown in the family unit, as male heads of households—or sometimes both parents—have left for the United States, leaving children behind with relatives, often grandparents. Children who have parents working abroad are especially vulnerable to community violence and forced migration as they can become targets for gang extortion due to the perceived or actual remittances they may receive. Additionally, as children enter teenage years and are increasingly at risk for victimization or recruitment by gangs, it becomes increasingly difficult for their relatives, especially elderly grandparents, to protect them. To this end, the United Nations Development Program reports that 26.7% of all inmates in El Salvador they interviewed in 2013 never knew their mother or father growing up. Schools no longer function as social institutions that offer a respite from the violence and instead have become de facto gang recruitment grounds. As a result of being targeted because of their family situation or perceived wealth, children flee, as a strategy to escape the gangs, to help support the family, and to reunify with their parents or other loved ones, many of whom have been separated for years.

c. Abuse in the home also has created stress, fear and motivation to leave the family home as well as the community. The pressure on families from local violence, economic uncertainty, and family member absence has a deleterious effect on the family unit, as instances of domestic abuse towards women and children have grown. It has been documented that more unaccompanied children are reporting instances of child abuse and neglect undertaken by non-parental caretakers. Children, in particular girls, are particularly exposed to domestic violence. A survey carried out by UNICEF revealed that 7 out of 10 unaccompanied children reported having been abused in their homes. In El Salvador it was reported that the domestic violence and sexual abuse of women and girls in the private sphere remain largely invisible and are consequently underreported.

d. Migrating children do not find the protection they need once they arrive in Mexico, even those who are eligible for asylum. The United Nations High Commissioner for Refugees (UNHCR) has consistently reported that an increasing number of unaccompanied children from Central America in particular are vulnerable to exploitation and cannot access protection in Mexico. To this end, UNHCR and USCCB are working with government authorities to provide training to law enforcement and protection officers on identifying and screening vulnerable children.

As an example of this lack of protection, USCCB found one children's shelter dedicated to caring for migrant children who may attempt an asylum claim in the Southern Mexico region, in Tapachula. Another shelter in Mexico City, run by the Mexican government's division of child welfare [Desarrollo Integral de la Familia (DIF)] houses children who have won asylum but cannot be released until they are 18.

Children who request asylum usually remain in detention for months, with little help to navigate the legal system. Once a child wins asylum, the only placement option available is the DIF child shelter in Mexico City until age 18, as there is no foster care system in place for these children. Shelter care is not intended to be a long-term placement for children, and often leaves children vulnerable to exploitation. Because of the challenges in gaining asylum in Mexico and the absence of an effective child welfare system, children often choose deportation back home so they can try to migrate again.

e. Countries of origin lack the capacity to protect children adequately. USCCB found that Guatemala, Honduras, and El Salvador lack the capacity to protect children in their law enforcement, child and social welfare, and educational systems. As mentioned, organized criminal networks and other criminal elements are active in many communities and schools, and the government is unable to curb their influence because of corruption, lack of political will, or lack of resources. Law enforcement personnel, low-paid and low-skilled, are compromised by these criminal elements. Child welfare services are virtually non-existent, as are foster-care and family reunification and reintegration services.

f. A significant number of migrants, particularly youth, have valid child protection claims. While the popular perception of many in the United States is that migrants come here for economic reasons, USCCB found that a growing number are fleeing violence in their homelands. UNHCR recently found 58% of the unaccompanied children it interviewed from Central America and

Mexico had some sort of international protection claim. A similar study in 2006 found only 13% of these children had a protection claim. Children who exhibit international protection concerns may be eligible to remain in the United States legally in some form of recognized legal status, such as Special Immigrant Juvenile Status, as an asylee.

U.S. Response to the Humanitarian Crisis

We support the [Obama] Administration's immediate response to this crisis, which created an inter-agency response led by the Federal Emergency Management Agency (FEMA). We offer the following recommendations to ensure that children are cared for throughout the legal process:

a. For the children, the faithful adherence to the best interest of the child standard is necessary in all decision-making. The best interest of the child principle is an internationally recognized child welfare standard used in the U.S. child welfare system. It refers to a process of determining services, care arrangements, caregivers, and placements best suited to meet a child's short-term and long-term needs and ensure safety permanency, and well-being. When applied in the United States special importance is given to family integrity, health, safety, protection of the child, and timely placement. This means that all procedures, protocols, and mechanisms developed are child-friendly, trauma-informed, and administered by child welfare professionals; that children are screened and assessed for their immediate humanitarian protection needs and their long-term international protection needs; that during the pursuit of long-term solutions for the children they are placed in the least-restrictive settings (i.e. community-based); that all children are connected with social and legal services to address their immediate needs; that long-term and durable solutions are pursued that are in the children's best interests; and that where repatriation is the best alternative available that safe repatriation and reintegration be conducted in collaboration and coordination with the children's home governments, NGOs, and other implementing partners.

Consistent with US child welfare norms, children should be placed in smaller community-based programs such as specialized foster care, group or small shelter programs which allow children to reside in family settings in communities. Large facilities are contrary to child welfare principles increase the risk of institutionalization, child maltreatment and losing track of children's individual needs.

b. For the United States government, a mutually supportive, intera-gency response is necessary to ensure we are leveraging the expertise and resources of the agencies that bear responsibility for addressing all aspects of the challenge. We are encouraged by the decision of the Administration to involve all relevant agencies of the government in responding to this crisis. The inter-agency work on the issue should incorporate clear leadership responsibilities and effective collaboration mechanisms to ensure the optimum results both in the United States and throughout the region.

c. Children should be properly screened and placed in the least restrictive setting, preferably with family or an appropriate sponsor. Children should be immediately screened, ideally by a child welfare specialist, as to whether 1) they are victims of human trafficking; and 2) whether they have special needs and require specific care, such as trafficking victims, children under 12, pregnant girls, and persons with disabilities. Where possible, children should be reunified with their family members during the course of their legal proceedings. Potential sponsors who can care for the child throughout the child's immigration proceedings should be identified and adequately screened. Children should not be released, pending fingerprint and background checks of their sponsors. HHS [Department of Health and Human Services] and other agencies should monitor, report, and respond to violations against children. As required under the law, expedited removal should not be used against unaccompanied children.

d. Families should be kept together, preferably in a community setting, and provided full due process rights. Families who are part of this migration flow, mainly women with young children, should not be detained in a restrictive setting. Alternatives to detention for these families should be explored, including with faith-based communities. Such models have been implemented in the past, with great success and at reasonable costs. The needs of mothers and children are best met in such a community setting, where their specialized needs can be met.

Moreover, subjecting these families to expedited removal procedures, as intended by the Administration, could undercut their due process rights. Many would be unable to obtain an attorney and, because of their trauma and the setting of the immigration proceedings, would be unable to adequately articulate their fear of return.

e. Post-release reception assistance should be expanded to meet the rising need. We urge increased post-release services which address family preservation, child safety, community integration, access to counsel and continued participation in immigration proceedings. The lack of sufficient funding for assistance post-release increases the likelihood of family breakdown, makes it more difficult for children to access public education and community services, and decreases the likelihood that the children will show up for their immigration proceedings.

With the release from custody happening on a shorter time frame—now less than 30 days—and with up to 90% of UACs [unaccompanied children] being released from ORR custody to communities, UAC resources need to be prioritized into community-based reception services which are located where families live. ORR could leverage the infrastructure and expertise of the U.S. resettlement agencies by providing all of the children community-based, reception services. Reception services should be required for all UAC to assist the family with navigating the complex educational, social service, and legal systems.

f. Pastoral care and services should be provided to children. Mr. Chairman, these vulnerable children should have access to pastoral services, including visitation by religious, including priests, minister, and other faith leaders. To date, requests for visitation to the border patrol stations and shelters for this purpose has been denied by the Border Patrol and ICE.

Conclusion

The situation of child migration from Central America is a complex one, with no easy answers. It is clear, however, that more must be done to address the root causes of this flight and to protect children and youth in the process. Clearly this problem is not going away; in fact, it is getting more urgent in terms of the dire humanitarian consequences.

Too often, and especially recently in the media, these children are being looked at with distrust and as capable adult actors, instead of as vulnerable and frightened children who have been introduced to the injustice and horror of the world at an early age. Anyone who hears the stories of these children would be moved, as they are victims fleeing violence and terror, not perpetrators. USCCB found that these children long not only for security, but also for a sense of belonging—to a family, a community, and a country. They are often unable to find this belonging in their home country and leave their homes as a last resort.

In conclusion, I ask you to consider the individual stories of these vulnerable child migrants and open your minds and hearts to their plight while seeking meaningful and long-term regional solutions. I ask you to respond to the needs of these children, not to turn them away or ostracize them, as Americans are a compassionate people.

MARK SEITZ is the Roman Catholic bishop of the Diocese of El Paso, Texas, and head of the Committee on Migration of the United States Conference of Catholic Bishops. He received an MA degree in theology from the University of Dallas.

EXPLORING THE ISSUE

Should the United States Deport Unauthorized and Unaccompanied Immigrant Children?

Critical Thinking and Reflection

1. Since almost all illegal and unaccompanied immigrant children are from Mexico, Central America, and other largely nonwhite countries, some charge that those want to deny them legal residency and eventual citizenship are racists. How do you evaluate this assertion?
2. Should families that arrive with minor children in the United States without authorization be treated the same or differently than unaccompanied children?
3. Is it reasonable to suppose that amnesty for illegal/unauthorized immigrants does not encourage further such immigration?
4. How much should the United States do financially and otherwise to follow Bishop Seitz's recommendations for easing poverty and violence in Mexico, Central America, and elsewhere?

Is There Common Ground?

Illegal/unauthorized immigrants, adults and children, are an ongoing problem in the United States that is not only not going to go away without significant changes but is also progressively getting more complex and even odd. Such immigrants are not a transient population. According to a Department of Homeland Security Study in 2011, 31 percent of these immigrants entered the country earlier than 1995. Another 55 percent had been in the country for between 5 and 15 years, with only 14 percent having arrived in 2005 or later. Moreover, there is a huge number, probably at least a million, of children and adults who were born in the United States of illegal/unauthorized immigrant parents. These children and adults are citizens, and deporting their parents after, in many cases, decades of residency would be harsh, indeed. Then there are the unaccompanied children. If they are allowed to stay, should their parents, siblings, and perhaps other members of their extended be allowed to join them?

At base, it seems that two things have to happen. One is that the flow of undocumented immigrants needs to be stopped. It makes a mockery of immigration law if those who cannot get into the country legally or who don't want to wait for entrance can try to enter anyway and, if they do so and are here long enough, can get drivers licenses, American schooling and other benefits, and even citizenship. It is also important to understand that to a degree, earlier undocumented immigrants were encouraged to come here or rewarded after they came by jobs and other incentives. Some had children when they came, some were parents whose children were born in the United States and thus citizens. Most have worked, paid taxes, and lived within the law, except for their illegal entry. Deporting them would be cruel and unusual punishment. There is something of a standoff, with Republicans wanting to halt the flow of illegal immigrants before creating ways for existing and clean-living undocumented immigrants to be able to come out of the shadows, get resident status, and progress toward citizenship. Doing both is what most Americans favor. A 2013 Pew Research Center survey found 77 percent of Americans favor a reform that combines allowing illegal/unauthorized immigrants who meet certain requirements to stay in the country providing for increased security along the borders. However, just like the advocates in this debate, Americans are split on their priorities. Some 49 percent say these immigrants should be permitted to gain legal status while border improvements are being made, and 43 percent want the path to residency to stay closed until after a secure border has been established. Moving ahead vigorously on both fronts would seem to be the common ground.

Create Central

www.mhhe.com/createcentral

Additional Resources

Ankarlo, Darrell. (2010). *Illegals: The Unacceptable Cost of America's Failure to Control Its Borders* (Thomas Nelson)

Bush, Jeb, and Bolick, Clint. (2013). *Immigration Wars: Forging an American Solution*. (Simon and Schuster)

Committee on Migration of the United States Conference of Catholic Bishops. (November 2013). *Mission to Central America: The Flight of Unaccompanied Children to the United States*, report of the Committee on Migration of the United States Conference of Catholic Bishops, headed by Bishop Mark Seitz. United States Conference of Catholic Bishops, Washington, D.C., on the web at: http://www.usccb.org/about/migration-and-refugee-services/

Nowrasteh, Alex. (June 8, 2014). "Unaccompanied Minors and Unintended Consequences." *The Hill* online, reproduced on the Cato Institute website at http://www.cato.org/publications/commentary/unaccompanied-minors-unintended-consequences?utm_source=feedburner&utm_medium=feed&utm_campaign=Feed%3A+CatoHomepageHeadlines+(Cato+Headlines)

Seghetti, Lisa, Siskin, Alison, and Wasem, Ruth Ellen. (July 28, 2014). *Unaccompanied Alien Children: An Overview*. Congressional Research Service, report R43599. On the web at http://research.jacsw.uic.edu/icwnn/files/2014/07/Unaccompanied-Alien-Children_An-Overview.pdf

Internet References . . .

Center for Immigration Studies

http://www.cis.org/

Immigration Policy Center

http://www.immigrationpolicy.org/

U.S. Census Bureau—Immigration

http://www.census.gov/population/intmigration/

U.S. Department of Homeland Security

http://www.dhs.gov/immigration-statistics

Selected, Edited, and with Issue Framing Material by:
John T. Rourke, *University of Connecticut, Storrs*

ISSUE

Was It Wise to Free Sergeant Bowe Bergdahl by Also Freeing Alleged Terrorists?

YES: Mark R. Jacobson, from "The Bergdahl Exchange: Implications for U.S. National Security and the Fight Against Terrorism," Testimony During Hearings before the Committee on Foreign Affairs, U.S. House of Representatives (2014)

NO: Ted Cruz, from "Guantanamo Bay Detainees: An Address to the U.S. Senate," *Congressional Record* (2014)

Learning Outcomes

After reading this issue, you will be able to:

- Understand the pros and cons of making deals with terrorist organizations.
- Be aware of the allegations regarding the capture of Sergeant Bowe Bergdahl that in the opinion of some impact this debate.
- Discuss the secretive process by which the deal was made and indicate whether you believe that it was a satisfactory process.

ISSUE SUMMARY

YES: Mark R. Jacobson, senior advisor to the Truman National Security Project, a private analysis and advocacy organization, supports the swap for Bergdahl on the grounds that to leave any soldier behind is not only unconscionable but would damage a sacred trust with our military personnel, leading some to question our nation's commitment to our troops, and could result in a tremendous propaganda victory for our enemies.

NO: Ted Cruz, a U.S. senator from Texas, criticizes the swap, arguing that the men and women of our military understand the value of protecting the national security of the United States of America, but are not comforted by negotiations with terrorists to release senior terrorist leaders who can once again begin actively waging war on the United States.

This much is certain. Twenty-year-old Sun Valley, Idaho, native Bowe Bergdahl joined the U.S. Army in 2008. He was deployed to Afghanistan in May 2009. The following month, on June 30, he disappeared from his post and was soon thereafter captured by the Muslim fighters. His captors released several video and audio recordings of Bergdahl beginning soon after he was captured in 2009 and in 2010 and 2012. During Bergdahl's captivity, the Army promoted him twice, first from private to specialist, then to sergeant. In May 2012, the U.S. government acknowledged that it had engaged in talks with the Taliban in an effort to free Bergdahl. U.S. officials told the press in early 2014 that contacts with the Taliban through intermediaries were again underway. Then on May 31, 2014, President Barack Obama announced that the Taliban had released the sergeant in exchange for five Taliban who were being detained at the U.S. naval base in Guantanamo Bay, Cuba, often referred to as Gitmo. After an extended medical checkup, Bergdahl arrived back in the United States on June 13 and after further medical review, resumed regular duty at the Fort Sam Houston military base in San Antonio, Texas.

One might assume that the return of a U.S. service-man to freedom would have been a time of national joy. But it was not to a substantial degree. There was significant criticism of the wisdom of the swap and the process by which it occurred. Also clouding the deal were allegations that Bergdahl was not a returned hero, but instead a deserter who had all but turned himself over to the enemy. Desertion from the military in time of war carries a potential death penalty.

Three days after Bergdahl's return, the Army indicated that it had appointed a major general to investigate the circumstances of Bergdahl's disappearance and capture. Soon thereafter, he announced that he had retained an attorney. As of this writing in September 2014, Sergeant Bergdahl remains on active duty and the investigation continues.

This issue has four levels of debate. The first debate relates to the wisdom of making deals with terrorists to free captives or achieve other ends. On one hand, the long-standing standard, "leave no soldier behind," means that every effort should be made to ensure all trapped or wounded soldiers are not abandoned and every effort should be made to secure the return of captured soldiers. This is said to be both a matter of honor and importance to the morale of the military.

Perhaps contradictory at times is the "no negotiations" with terrorists standard. This dates at least back to pronouncements by President Ronald Reagan (1981–1989). It was reiterated as recently as 2013 after terrorists in Algeria took about 30 prisoners, including some Americans. When a reporter asked a U.S. State Department official about rumors of a potential trade of Sheikh Omar Abdel-Rahman, mastermind of the World Trade Center attack in 1993, for the hostages in Algeria, a department spokesperson replied, "The United States does not negotiate with terrorists." The logic behind negotiating with terrorists is that it provides recognition, ransom, the freeing of captured terrorists, and other positive rewards that encourage terrorists everywhere to commit further abductions. Whatever the logic, it is also the case that countries, including the United States, have not strictly complied with the "no negotiations" dictum. Indeed, President Reagan worked with Israel when it freed 700 Arab prisoners in return for several Americans being held from a hijacked TWA flight. Under President George Bush in 2007, the United States cooperated with the British in a deal that released an Iraqi Shiite radical prisoner in exchange for the release of a British civilian being held by the Shiites.

When after the Bergdahl swap a State Department spokesperson was asked about the "no negotiate" standard, she replied, "Well, our line is that we don't make concessions. I mean, that's the—you're quoting it colloquial. That's actually not what you'll hear us say from the podium." She went on to justify the negotiation as akin to those between governments during a war, saying the swap was "consistent absolutely with what's happened in previous wars, including Korea, including Vietnam." This point did not note that in 2010 the State Department designated the Taliban group, Tehrik-e Taliban Pakistan, as a terrorist organization. When a reporter asked yet another State Department official whether the Taliban was a terrorist group, he demurred, saying, "We regard the Taliban as an enemy combatant." It may be that Bergdahl was not being held by another Taliban group or even by the "Haqqani network," an Islamist insurgent group allied with the Taliban and al Qaeda. An irony is that the network reportedly began in Afghanistan in the 1970s and was aided by U.S. intelligence agencies in their effort to disrupt the then Soviet occupation of Afghanistan.

The second debate in this issue relates to the process and whether the president should have involved Congress and even the public in the decision to swap terrorists for Bergdahl. Although, as noted, there were some public acknowledgments by the Obama administration that it was seeking Bergdahl's release, the negotiations were conducted in secret. Proponents would argue that public negotiations would have doomed the talks to failure. Critics would say that decision was important enough to merit a national debate including Congress and the public. Critics say that was particularly so given the allegations against Bergdahl.

The third debate here is whether the deal was a good one. The five Taliban, all of whom had been held at Guantanamo for 12 years, are Khairullah Khairkhwa, the governor of Afghanistan's Herat province; Mullah Norullah Noori, governor of Balkh province and senior Taliban commander in the northern city of Mazar-e-Sharif when the Taliban fought U.S. forces in late 2001; Mohammad Fazl, army chief of staff under the Taliban regime; Abdul Haq Wasiq, deputy chief of the Taliban regime's intelligence service; and Mohammed Nabi, chief of security for the Taliban in the city of Qalat. The five were released to Qatar, a small emirate on the Persian Gulf, where they must remain for one year under Qatari supervision. The issue is whether these individuals were too dangerous to release. After one year, that is, in June 2015, the five will be free to leave Qatar and, if they choose, return to the Taliban in Pakistan or Afghanistan.

The fourth debate within this issue involves whether Sergeant Bergdahl has any culpability for his capture and, if so, whether that impacts the value of the swap that freed him. He wrote to his parents saying he was disillusioned with the war, but some of those who served with

Bergdahl accuse him of being two-faced. "Bergdahl was complaining to his parents that our platoon was committing atrocities instead of helping the local populace," one fellow soldier testified to the House Foreign Affairs Committee, "but he was telling our platoon that we needed to stop trying to win hearts and minds and focus more on killing the Taliban." Also a mystery to probably everyone except Bergdahl is how he was captured. He claims that he fell behind his platoon on patrol and was captured. Others say Bergdahl is a deserter. One member of Bergdahl's unit says there was no patrol the night he disappeared and charges that he voluntarily left his unit in the face of the enemy. There are also allegations that at least six fellow soldiers were later killed in the military's subsequent effort to "leave no soldier behind" by finding Bergdahl.

In the first of the following selections, Mark R. Jacobson argues that the Taliban-for-Bergdahl swap was the right thing to do whatever else the specific circumstances were. Senator Ted Cruz disagrees. He criticizes both how the deal was made and that it was made at all.

YES

Mark R. Jacobson

The Bergdahl Exchange: Implications for U.S. National Security and the Fight Against Terrorism

As someone who was in the Pentagon on September 11, 2001, the threat posed by terrorism is not lost on me. While I made the decision years before to devote myself to serving our nation, that day changed many of our lives forever. As a result I spent several years in Afghanistan—some as a Naval intelligence officer and some as a civilian advisor—and I am acutely aware of the danger that remains to Afghanistan. For all of us at the table the conflict in Afghanistan is personal and we all feel the impact of this war in a way most Americans do not.

I have been asked to address several issues pertinent to [this committee's] national security oversight responsibilities related to the potential risks incurred by exchanging SGT [sergeant] Bowe Bergdahl for five Afghan detainees held at Guantanamo Bay, Cuba and the potential precedent set by negotiating with terrorists or insurgents. In short, I believe that securing SGT Bergdahl's release was absolutely the right thing to do and was worth the potential risks. One of the greatest commitments an American can make to their nation is to put on a uniform and take an oath as a member of the U.S. armed forces to "support and defend the Constitution of the United States."

By taking this oath, these men and women—who are sons and daughters, fathers and mothers—make the selfless decision to put their country first. They do so knowing that they may one day be called to give what Abraham Lincoln called the "last full measure of devotion,"—to give their lives for their comrades, their families, and their nation.

With each of these volunteers, the military makes its own promise to be there for those who have been captured. The commitment is simple: leave no man or woman behind; no exceptions. This obligation is something we owe to all who have served, are serving, and will serve. Some might suggest that we should not have risked lives or time to find and retrieve Bergdahl because of the potential circumstances surrounding his capture.

But this commitment to our captured soldiers is unequivocal and must take place regardless of the circumstances of their capture. The Chief of Staff of the Army, General Raymond Odierno, has promised a thorough and transparent investigation into the circumstances surrounding Bergdahl's capture, and the Chairman of the Joint Chiefs of Staff, General Martin Dempsey, has been clear that that leadership "will not look away from misconduct if it occurred." But the obligation to retrieve SGT Bergdahl and the circumstances of his capture should not be conflated. Indeed, if Bergdahl did act improperly, then it is even more important that he be held accountable in the military system for his actions. To leave any soldier behind is not only unconscionable but would damage a sacred trust with our military personnel, lead some to question our nation's commitment to our troops, and could result in a tremendous propaganda victory for our enemies. Additionally, given the ongoing transition in Afghanistan it was better to do this deal now while we have military leverage. Indeed, there was also a tremendous risk of having a captured U.S. soldier being executed on video as happened with American businessman Nick Berg Iraq in 2004 and journalist Daniel Pearl in Pakistan in 2002—acts which not only demonstrated the brutality of our enemy but could be used to foster the recruitment of extremists and spread extremism.

Some argue that the release of 5 detainees from Guantanamo is itself a propaganda victory. In the big picture, the effects of a Taliban propaganda campaign will be short-lived and pale in comparison to recent strategic changes in Afghanistan to include a transition to Afghan security's "self-reliance" and two successful rounds of elections, each of which saw around 7 million Afghans vote—40% of them women—in open defiance of the Taliban.

While there will always be some risk posed by the release of detainees from Guantanamo Bay, these risks also held true during the Bush Administration when

Mark R. Jacobson. Testimony during hearings on "The Bergdahl Exchange: Implications for U.S. National Security and the Fight Against Terrorism" before the House Committee on Foreign Affairs, June 18, 2014.

532 detainees were released from Guantanamo Bay between 2002–2009, some of whom have returned to the fight. Despite the potential risks of releasing detainees from Guantanamo, there are several reasons why we should temper our concerns:

First, these detainees will be held by Qatar for the next year and will be subject to specific security measures to limit their activities and potential to become threats. As Secretary of Defense [Chuck] Hagel testified to the House Armed Services Committee [HASC] on June 11th of this year, the Qatari government recently signed a Memoranda of Understanding with the United States that included "specific risk mitigation measures and commitments from the Government of Qatar [including] travel restrictions, monitoring, information sharing and limitations on activities, as well as other significant measures" that the Department detailed in a closed portion of the Armed Services Committee hearing. As the Secretary noted, President Obama also received a personal commitment from the Emir of Qatar to uphold and enforce the security agreements outlined in this agreement.

Second, there is not a consensus that these five individuals will inevitably return to the battlefield. A quick review of publicly available materials demonstrates that recidivism is not a certainty by any means, with a rate hovering at around 10%. Statistics provided in the September 2013 unclassified summary of the "Reengagement of Detainees Formerly Held at Guantanamo Bay, Cuba" report provided by the Director of National Intelligence [DNI] to Congress note that the number of detainees confirmed in re-engaging on the battlefield had been about 16.6%. A closer look reveals that of the 532 detainees released before January 22, 2009, the "confirmed" recidivism rate was around 18% but of the 71 individuals released since 2009, only 4.2% were confirmed as having returned to the battlefield. A similar drop in rates has been seen with those who are suspected, but not confirmed, to have re-engaged on the battlefield. According to the DNI figures, the United States has proven more successful in the past six years than during the time between 2002–2009 in reducing recidivism rates from about 30% to about 10%. Reporting by terrorism expert Peter Bergen at the New America Foundation reinforces the notion that recidivism rates are probably even lower at around 8.7%.

Third, even in the event that they do return to the battlefield, the Afghanistan of 2015—even 2014—is not the Afghanistan from which they were captured. Average life expectancy in Afghanistan is now 60 years vs. 42 years in 2001. At the time of the fall of the Taliban, just under a million children went to school, nearly all male. Now over 9 million children are in Afghan schools, nearly 40% of them girls. Most importantly, there is new hope—especially amongst the youth—that they can live in an Afghanistan that is at peace. Taliban insurgent networks are shattered in many places, the Afghan National Security Forces are much more capable fighters, and as already noted, the political situation in Afghanistan demonstrates that for all the Taliban's efforts, they cannot stop a peaceful transition of civilian power. The strength of the Taliban will not return because of the presence of these five individuals who have been off the battlefield for over a decade, and I am not entirely certain that they will be welcomed with open arms by their former colleagues who may not trust the fact that these individuals have been with the Americans for so long.

Furthermore, it is worth considering the potential opportunities that have been created by the completion of this exchange. Qatar has already proven an acceptable "neutral" location for the Taliban to send their representatives in search of an eventual peace within Afghanistan. We should continue to work with the Afghan government to leverage Qatari credibility to help move talks towards an eventual peace agreement. Indeed, retired Marine Corps General James Mattis has even suggested a military advantage to the exchange, noting that Bergdahl's release has created a "military vulnerability" for the Taliban and the Haqqani network [an Islamist group led by Maulvi Jalaluddin Haqqani and allied with the Taliban]. In short, there is now freedom for the U.S. to operate against them now that they no longer hold a U.S. prisoner.

All of us would like to see an end to the conflict in Afghanistan but this will bring with it questions about the final disposition of detainees still held at Guantanamo Bay. Part of this will have to do with how lawyers define the end of "hostilities" in Afghanistan. While I cannot speak to it as a legal issue, from a political perspective it is hard to envision any comprehensive peace agreement between the Government of the Islamic Republic of Afghanistan and the Taliban that does not involve the return of the remaining Afghans at Guantanamo to Afghan government control. Therefore, it would be wise for us to generate as much political value out of the Afghan detainees while we have them. Arguably this is what was done with the five recently released. As Secretary Hagel noted in his HASC testimony, none of those detainees had been "implicated in any attacks against the United States, and we had no basis to prosecute them in a federal court or military commission."

The alternative is to keep them in confinement forever without any charges.

Additionally, as history has shown, an exchange of prisoners does not have to wait until after a war ends but can happen as wars draw to a close, as part of potential or actual negotiations and before the final armistice or peace-treaties are completed. For example, while the Korean War armistice was not signed until the end of July 1953, both sides had already conducted Operation "Little Switch," (April 20–May 3, 1953) where 684 U.N. sick and wounded troops (including 149 Americans) were exchanged for 1,030 Chinese and 5,194 Korean troops. Indeed, major fighting continued after this exchange of prisoners including the Battle of Pork Chop Hill, engagements in the Kumsong River Salient, as well as some of the largest U.S. Navy and Marine Corps air operations of the war. Even during the Second World War, there is a record of at least one exchange prior to the conclusion of hostilities, in this case, in November 1944, when A. Gerow Hodges, an International Red Cross worker detailed to the U.S. 94th Division, was able to convince German military authorities to swap 149 American POWs for a like number of German prisoners.

Some question whether the recent prisoner exchange created a precedent that will engager the lives of U.S. personnel and has broken from past practice of not negotiating with terrorists. I think this assessment is too simple and in some cases disregards historical precedent. First, the threat of kidnapping to U.S. members of the armed forces, diplomats, and citizens has long been the case in Afghanistan, and our forces have been prepared for that throughout over a dozen years of conflict in Afghanistan. Indeed, I felt this was my own greatest threat during my military and civilian service in Afghanistan. There is no reason to think that this calculus will be altered by the recent exchange. In short, terrorists and insurgents with whom we are at war have wanted to kidnap Americans before and will most certainly keep trying to in the future.

The deal to retrieve SGT Bergdahl was a prisoner exchange, not a negotiation with terrorists. But that said it is important to note that the popular view that the United States does not negotiate with insurgents, terrorists, or even state sponsors of terrorism is not historically accurate. In terms of the Taliban alone, the United States has been talking and negotiating with the Taliban for some time in recognition that the war in Afghanistan cannot end without a political settlement.

A quick review of history illustrates that at particular times, the United States has found it necessary to negotiate with terrorists and state-sponsors of terrorism. In 1968, the United States negotiated with the North Korean government in 1968 to obtain the release of 83 American personnel on the USS *Pueblo* that had been boarded and captured by the North Koreans.

As former State Department official Mitchell Reiss has noted in his book *Negotiating with Evil*, President Nixon pressured allies, including Israel, to release prisoners as part of negotiations with the Popular Front for the Liberation of Palestine in order to resolve the hijacking of two hijacked airliners; the Iran hostage crisis of 1979–1981 was resolved in part by the agreement to unfreeze $8 billion in frozen Iranian assets; and of course, there was the "arms for hostages" deal negotiated by the Reagan Administration as part of what eventually become the Iran-Contra affair.

Likewise, while the negotiations were rather one-sided, Ambassador Robert Oakley did meet with the late Somali warlord, Mohammad Farah Aideed to secure the release of Chief Warrant Officer Michael Durant who was held in captivity after the Battle of Mogadishu in October 1993.

Other nations have done this as well—Margaret Thatcher negotiated secretly with the Irish Republican Army and while Israel has at times famously said it will not negotiate with terrorists, we know that successive Israeli administrations have made prisoner exchanges—at times trading a thousand prisoners for just a few Israeli soldiers. But in all these cases it is important to distinguish between those situations more akin to what is expected in war—e.g. a prisoner exchange, part of a complex series of counterinsurgency initiatives. In other words, negotiating with a terrorist group or a state-sponsor of terrorism does not necessarily equate to paying ransom for hostages.

I understand the disappointment we feel in the stories coming out about Sergeant Bowe Bergdahl, and I understand the anger felt by some of his comrades who feel that he deliberately left his post. If I were them, I might feel the same way. But the truth is that we do not know the truth. Unfortunately, the process to determine it is impacted by all the speculation in a public setting. In our nation of laws the presumption of innocence is sacrosanct, an age old principle that demands, even if we believe with all our being otherwise, that people are innocent until proven guilty. Now that the Department of Defense has announced its intent to have Maj. General Dahl lead an investigation of the facts and circumstances surrounding Sergeant Berghdal's disappearance and capture, it is imperative to preserve the integrity of that investigation—it must be thorough and allowed to take place without politics or partisanship. Without it we are unlikely to ever have accountability.

We may not like it, but in the end, foreign affairs and national security policy are often about juggling bad options and finding the least worst of these options; there are rarely simple solutions. The decision to exchange Sergeant Bergdahl may be imperfect, but in my mind, it represented the right approach to balancing national security needs and does not in any way prevent the United States from continuing to prosecute a war with our Afghan partners against the Taliban nor does it appreciably increase the risk of new threats. We have been negotiating with the Taliban to find a solution in Afghanistan and we have precedents for negotiating with groups such as the Taliban for prisoner exchanges. The potential threat posed by these detainees must be looked at within the context of the Afghanistan to which they will return. Regardless, we never leave our soldiers behind.

MARK R. JACOBSON is s senior advisor to the Truman National Security Project, a private analysis and advocacy organization. He has a PhD degree in military history and strategic studies from Ohio State University.

Ted Cruz

Guantanamo Bay Detainees: An Address to the U.S. Senate

I rise [in the Senate] today to raise an issue that has been of growing concern to the American people: the exchange of the so-called Taliban five—five terrorist detainees from Guantanamo—in exchange for Sergeant Bowe Bergdahl.

Let me say from the outset, this is not about Sergeant Bergdahl. The circumstances under which he became a prisoner of the Taliban is an issue for the Army. There was an investigation into this matter in 2010, and hopefully the Army will be able to bring clarity to that situation soon. What I wish to speak about today is keeping the American people safe from the terrorists who attacked us on September 11, 2001, resulting in the deaths of 2,977 innocent people.

The Taliban five are among the worst of the worst. They were all high-level officials in the Taliban regime who gave aid and support to Al Qaeda in Afghanistan in the period leading up to the 9/11 attacks. These five were designated "high" risk by the Guantanamo Review Task Force convened in 2009 on the orders of President Obama, whose report was published on January 22, 2010. Two of the five are wanted by the United Nations for war crimes against Afghan civilians.

Khairullah Khairkhwa, for example, was described in his GTMO case file as "a hard-liner in Taliban philosophy" with "close ties to Osama bin Laden." Mohammad Fazl was second in command of the Taliban army in 2001. These were not junior-level players. [Guantanamo Bay Naval Base is referred to Gitmo or GTMO.]

Capturing these five men was a priority when our troops participated in the liberation of Afghanistan from the Taliban in 2001, where our sons and daughters bled and died to free Afghanistan and to exact punishment on those who carried out a horrific terrorist attack on the United States of America. We cannot know for sure how many American soldiers paid the ultimate price to capture these five senior terrorists.

Even as many other detainees at GTMO have been released, up until now, these five have been considered too dangerous to let go. Given the level of threat they represent, any proposal to release them should be of the utmost seriousness. Unfortunately, by all indications the administration's release treated their threat as anything but serious.

Americans need to know how the Obama administration thinks it has made our nation safer by negotiating with terrorists to release these five dangerous terrorist leaders. Until President [Barack] Obama can make his case and convince the American public that this swap was in our national interests, prudence dictates that all further transfers and releases from Guantanamo Bay should be off the table.

Unfortunately, there have been no answers from this administration on how this deal furthers the national security interests of the American people or why the deal was so urgent that the administration refused to comply with its legal obligation to inform Congress 30 days before the transfer. Instead, the administration has vilified those who would raise questions about it as somehow not being concerned about securing the return of our troops. That attack—that slur—shouldn't even be dignified by a response, particularly given what has been publicly admitted. President Obama has publicly admitted that there is "absolutely" a chance of the Taliban five returning to the battlefield and attacking Americans.

Indeed, the current Taliban leadership has announced that from their perspective this deal is so good for them that they should now prioritize kidnapping other Americans. For example, last Thursday one top Taliban commander told *Time* magazine—and this is a quote—"It's better to kidnap one person like Bergdahl than kidnapping hundreds of useless people. It has encouraged our people. Now everybody will work hard to capture such an important bird." This deal puts every soldier, sailor, airman, and marine—every man and woman standing up to defend this nation—in jeopardy.

The chair of the Senate Intelligence Committee, Senator Dianne Feinstein [D-CA], has publicly said that

Ted Cruz. Address to the U.S. Senate, June 9, 2014.

she has seen "no evidence" that Sergeant Bergdahl was under urgent threat in recent weeks or months.

All of these admissions together raise serious and legitimate concerns about the circumstances of the release of the Taliban Five, and they also make clear that the administration should stop vilifying any who raise these national concerns. Instead, the President should stand up and honor his commitment to the American people, defend this decision in terms of the national security interests of the United States—what should be the highest priority for the Commander in Chief.

Instead, we have recently learned from news reports that there are at least four other Gitmo detainees who are being considered for release. So not only has there not been accountability as to why this happened, but it appears the administration wants to go down the same road and I can only assume is willing again to violate the law and not notify Congress the next time, just the way it violated the law by not notifying Congress this time.

Before any further such action is considered, we need to take a pause and assess what happened with the Taliban five. We need to answer:

- Who did the vetting that resulted in the assessment that the Taliban five no longer posed a high level of threat to the United States?
- Who participated in the decision to release them?
- Was this the same deal the administration says they offered to brief Congress on previously or is it something different?
- Was the President fully briefed on the background of the Taliban Five and the likelihood of recidivism?
- How did the administration reach its apparently high level of confidence that the Taliban five will be secure in Qatar?
- How did they arrive upon the notion that that security should last only 1 year, after which the American people will be safe if these terrorists are released altogether? On what basis did the administration judge that only 1 year was sufficient?
- How was the decision made to ignore the law and bypass Congress, including bypassing the chairs of the Senate and House Intelligence Committees, Foreign Relations Committees, and Armed Services Committees?
- In what circumstances does the administration intend once again to openly defy the law and refuse to provide notification to Congress?

These are questions, I might note, that should be bipartisan concerns. This should not be a partisan affair— asking questions that affect the national security of every single American citizen and every single man and woman serving in the military.

In order to give the Obama administration the opportunity to satisfy the many outstanding questions the American people have about their safety—and I would note, having just returned from Texas, I found over and over again Texans, men and women, asking these very questions—I will propose this week that before we consider any additional releases from Guantanamo, we answer these questions first.

The legislation I will be filing, No. 1, will immediately call for a 6-month freeze on any Federal Government funding to transfer detainees from Guantanamo. No. 2, to enforce this requirement, the legislation will provide that, should the President choose to disregard this law—as, sadly, has been his pattern so many other times—all funds expended in the transfer would be deducted directly from the budget of the Executive Office of the President. No. 3, because we understand that conditions might possibly arise that would necessitate the release of an individual prisoner and out of respect for the President's special role in international matters, this legislation explicitly provides a means for the President to ask Congress for a waiver of the 6-month bar in an individual case. But, finally, because we believe the release of detainees from Guantanamo—which holds some of the most dangerous people on the planet— is a matter of the gravest import, this legislation would require that for every order for release of a Guantanamo detainee, it must be personally approved by the President. This would ensure that the fullest consideration and deliberation goes into the process.

This latest deal—which was announced to the American people as a fait accompli, with no opportunity for Congress to scrutinize it, no opportunity for the American people to assess it—this latest deal constituted negotiating with terrorists to release five senior terrorist leaders, and it raises obvious questions.

First of all, how many Americans did these five terrorist leaders directly or indirectly murder? How many lives—American lives—are they responsible for taking?

Second, how many American soldiers gave their lives to capture these five senior terrorist leaders? How many graves do we have of sons and daughters of Americans because they were sent in to capture these five who have just been released?

Third, given their release—and the President's admission that there is "absolutely" a chance that they will return to actively waging war against the United States—how many Americans are at risk of being killed directly or indirectly by these terrorist leaders we have just let go?

Finally, if the Taliban five do return to actively trying to kill Americans, how many American soldiers will once again have to risk their lives or, indeed, will give their lives trying to kill or capture these terrorists once again?

These are questions of the utmost seriousness, and to date the administration has not even attempted to answer them. Instead, it has suggested that anyone raising these questions is simply failing to stand by the men and women of our military. I can tell you, the men and women of our military understand the value of protecting the national security of the United States of America, and the men and women of our military are not comforted by negotiations with terrorists to release senior terrorist leaders who can once again begin actively waging war on the United States.

Every American is naturally eager to end the long war in Afghanistan, but that does not mean we disregard the threat that violent terrorist groups such as the Taliban pose to our Nation. We know from the hard experience of the last decade that at least one in three Guantanamo detainees has returned to the battlefield. That has been what history has taught us.

Until we have full confidence that this threat to American lives is being fully and properly assessed, that we are taking steps to protect the lives of American civilians and American soldiers and sailors and airmen and marines, it is only prudent to take the steps in the legislation I am introducing this week, and I hope the Senate will do so.

TED CRUZ is a Republican U.S. senator from Texas. He has a JD degree from Harvard University.

EXPLORING THE ISSUE

Was It Wise to Free Sergeant Bowe Bergdahl by Also Freeing Alleged Terrorists?

Critical Thinking and Reflection

1. In a situation where the two standards, "leave no soldier behind" and "no negotiations with terrorists," are in conflict, which should prevail?
2. Do you agree with Mark Jacobson that trading the Taliban five for Sergeant Bergdahl was appropriate, no matter what the circumstances of his capture were?
3. Assume that Bergdahl did voluntarily desert his post, a crime under military law. But also assume that he was experiencing tremendous stress. Among other things, his unit had lost its first member who Bergdahl knew when a lieutenant with whom Bergdahl was serving was killed in action. Should factors such as post-traumatic stress disorder (PSTD) mitigate or even negate any charge of desertion or dereliction of duty?
4. It is almost certain that by the time you read this, the military will have completed its investigation of the circumstances surrounding Bergdahl's disappearance and issued a finding. Find out that result, if you do not already know. From what you know, do you agree with it? Does it change your view of the value of the trade?

Is There Common Ground?

Finding common ground for each of the four questions that are part of this overall issue is difficult. First, should the United States negotiate with terrorists? The obvious middle ground is "sometimes." But is that really a policy that will deter terrorists? And if the answer is "sometimes," then the question is, "when?" Does the United States only negotiate to free capture American military personnel? How about American diplomats and other government officials? Do they merit more protection than any American taken captive by anyone, anywhere?

The second issue, the secrecy of the negotiations, is also tricky. President Woodrow Wilson (1913–1921) long ago in his Fourteen Points call for open treaties, openly negotiated, but that standard, especially the openly negotiated part, has been routinely ignored by most government including that of the United States. The problem with open negotiations is that diplomacy is vastly more difficult amid a cacophony of criticism and nit picking. Yet democracy requires public awareness and input into the policy process.

The third issue is whether the deal was a good one. That requires weighing the value of Sergeant Bergdahl versus that of the five Taliban. From what is known the five were not terrorists in the most immediate sense. They were more part of the Taliban regime in Afghanistan that the U.S. toppled. However, the Taliban were supporters and facilitators of al Qaeda terrorism. It also seems reasonable to guess

that after the year's stay in Qatar they will return home and once again become Taliban leaders. It may never be known, but maybe there was some compromise here, with the U.S. refusing to return even more dangerous individuals or limiting the number who were swapped. This fourth issue also involves the value of Bergdahl and whether that was/should be affected by whether he deserted or was honorably captured, and, if he did desert, should any PSTD he was suffering be considered? You can argue that it is impossible to set a value on a Bergdahl or any other individual, but countries do that implicitly all the time, especially when they go to war and implicitly accept some deaths and injuries in exchange for accomplishing a goal.

Create Central

www.mhhe.com/createcentral

Additional Resources

Neumann, Paul R. (2007). "Negotiating with Terrorists." *Foreign Affairs* (86/1: 128–138)

Phillips, Peter J., and Pohl, Gabriela. (2013). "Does Negotiating with Terrorists Make Them More Risk Seeking?" *Journal of Politics and Law* (6/4:108–27).

Schlueter, David A. (2013.) "The Military Justice Conundrum: Justice or Discipline?" *Military Law Review* (215:4–73)

Internet References . . .

U.S. House of Representatives, Committee on Armed Services

http://armedservices.house.gov/index.cfm/hearings
-display?ContentRecord_id=C938E2BF-6136-4FE4
-BFBE-E3EFEC9C7025&ContentType_id=14F995B9
-DFA5-407A-9D35-56CC7152A7ED&Group
_id=41030bc2-0d05-4138-841f-90b0fbaa0f88

U.S. House of Representatives, Committee on Foreign Affairs

http://foreignaffairs.house.gov/hearing/joint
-subcommittee-hearing-bergdahl-exchange
-implications-us-national-security-and-fight

U.S. Department of Defense

A good site to follow the story of Bowe Bergdahl from the official point of view is the news site of the Department of Defense. Keyboard his name into the search window.

http://www.defense.gov/

ISSUE

Selected, Edited, and with Issue Framing Material by:
John T. Rourke, *University of Connecticut, Storrs*

Has Edward Snowden Damaged the U.S. Public Interest?

YES: **William H. Harwood,** from "Whistleblower or Traitor, Snowden Must Shut Up," posted to the *Huffington Post* (2013)

NO: **Emily Bazelon,** from "Is Edward Snowden a Traitor? If He Is, So Was Daniel Ellsberg," *Slate* website (2013)

Learning Outcomes

After reading this issue, you should be able to:

- Appreciate the tension between whistle-blowing and national security in a democracy.
- Understand why some Americans can describe Edward Snowden as a hero and other Americans condemn him as a traitor.
- Assess the difficulties of protecting secrets in the electronic age.
- Discuss the damage that leaked information can cause to U.S. foreign relations and national security.
- Weigh whether it is better for Americans to know information even if its disclosure may harm foreign relations and national security.

ISSUE SUMMARY

YES: William H. Harwood, an instructor of philosophy, the University of Texas–Pan American, argues that whatever Edward Snowden's motivations were, his leaking of classified national security information was wrong because it damaged U.S. foreign policy.

NO: Emily Bazelon, a senior editor at *Slate*, an online political journal, writes that while the issue of divulging national security secrets is a difficult one, Americans have a right to know what their government is doing, and Edward Snowden's leaks did more to advance people's democratic right to know what their government is doing than to damage U.S. foreign policy.

In 2012, Edward Snowden went from an obscure information technology specialist with the U.S. National Security Agency (NSA) to a famous, many say infamous, global figure. What propelled Snowden to celebrity was his decision to leak a huge volume of classified U.S. national security and foreign policy documents to the press in the United States, Great Britain, and elsewhere. Some of documents embarrassed the U.S. government. For example, the 200,000 or so documents included information showing that the United States had spied on the governments of such friendly coun-

tries as Australia, Brazil, France, Germany, Great Britain, Israel, Mexico, Norway, and Spain. Snowden's disclosures also made it clear the NSA had extensively monitored internal U.S. communications among citizens and organizations by monitoring phone records and hacking into the communications networks of Google, Yahoo, and other Internet service providers. Estimates of the number of intercepts kept as a record by NSA run into the millions per day. It remains unclear how much of this activity was known to President Barack Obama and other top administration officials and whether all the NSA's activity occurred lawfully.

What is clear is that Snowden's revelations set off a storm of controversy. Numerous countries and their leaders protested being spied upon. For one, Germany's Chancellor Angela Merkel personally called President Obama in response to reports that the NSA had intercepted her cellphone conversations. In response, the Obama White House issued a statement that it was not and would not do so, but did not say that it had not done so.

Reactions in the United States to what Snowden had done were widespread. Snowden characterizes himself as a "whistle-blower," who made public illegal U.S. government activity. As he put it, "All I wanted was for the [American] public to be able to have a say in how they are governed." Speaker of the U.S. House of Representatives John A. Boehner condemned Snowden as a "traitor," charging that, "The disclosure of this information puts Americans at risk. It shows our adversaries what our capabilities are. And it's a giant violation of the law." Taking a diametrically different stand, Steve Wozniak, cofounder of Apple, depicted Snowden as a "hero to my beliefs about how the Constitution should work. I don't think the NSA has done one thing valuable for us . . . that couldn't have been done by following the Constitution." Many U.S. leaders took a somewhat nuanced position. President Obama said, "I don't think Mr. Snowden was a patriot," but the president also announced a review of the NSA's practices, thereby implicitly acknowledging that the disclosures were both grounded in fact and troublesome. Yet others defended the NSA's activities. Its director, General Keith Alexander, credited the NSA's intelligence-gathering programs with having prevented 54 terrorist attacks globally, including 13 in the United States.

As for the American public, it was ambivalent about Snowden. When asked in a July 2013 national poll whether Snowden was a whistl-eblower or a traitor, 55 percent said he was a whistle-blower, while only 25 percent replied that he was a traitor. Yet, in a nearly direct reversal of the numbers, another national poll that month found that 56 percent of Americans disapproved of Snowden's actions, with just 25 percent approving.

Snowden's actions somewhat paralleled those of U.S. Army Private Bradley Manning, an analyst with Army Intelligence. In 2010 he copied 750,000 pages of classified documents and videos about the wars in Iraq and Afghanistan and sensitive correspondence written by U.S. diplomats and leaked them to WikiLeaks. That information was also analyzed and reported by news outlets around the world. Manning was court-martialed in 2013. A military judge acquitted Manning of aiding the enemy, but convicted him of violating the Espionage Act and sentenced him to 35 years in prison.

Snowden was luckier. In June 2013 the U.S. government also charged him with violating the Espionage Act, but prior to that Snowden had fled the country to Hong Kong. He applied for political asylum to many countries, but most of them, perhaps because of U.S. pressure, refused him entry. Russia did take him in, where he remains.

In the first of the two selections that follow, University of Texas–Pan American instructor William H. Harwood neither calls Snowden a hero nor a traitor, but does argue that he has done more harm than good. In the second selection, journalist Emily Bazelon portrays Snowden as a whistle-blower. To buttress her case, she cites the example of Daniel Ellsberg. In 1969 he leaked an extensive collection of classified U.S. documents to the *New York Times* and other newspapers. In part, they showed that the U.S. government was not telling the public the truth or the whole truth about its motivations, actions, and goals in the Vietnam War. These documents became known as the Pentagon Papers. The government indicted Ellsberg, but a federal judge dismissed the indictment because of government misbehavior during the investigation. The government also tried to prevent publication of the Pentagon Papers as a violation of the Espionage Act. Eventually, the Supreme Court ruled in the landmark case, *New York Times v. U.S.* (1971), that newspapers have the constitutional right under the First Amendment to publish the documents.

YES ⬅

William H. Harwood

Whistleblower or Traitor, Snowden Must Shut Up

Everyone keeps debating whether Edward Snowden is a patriot or a traitor. But his recent revelations, describing in tantalizing detail the spying on 70 million French phone calls or hacking the phones of Mexican presidents—remove all doubt on one thing: Snowden has got to shut up.

Since his bombshell revelations first hit in June [2013], the country—and the pundits—have not really known what to make of Edward Snowden. Is he a whistleblower or a traitor? In a rare moment of bipartisanship, those in D.C. seem to agree on the latter. . . . The pundits are torn—some even with themselves. The liberal blogosphere has all but accepted his divinity. Public opinion has remained pendulous.

This really isn't hard. All intelligence officers, like all uniformed personnel, federal employees, presidents, etc., must take an oath to protect the U.S. Constitution from enemies both foreign and domestic. Once upon a time Snowden could have been a whistleblower, patriot, hero. Alas, now he is a traitor—if nothing else, against himself and the American interests which he claims to be defending.

From the beginning Snowden's story was incoherent. A kid in his twenties—not a career or seasoned employee—is offered access to seemingly infinite amounts of classified material and somehow absconds with such unnoticed to foreign soil on a whim.

(Incidentally, *this* is the real story: how the outsourcing of our intelligence community led to such a galactic failure in the national security of the most powerful nation in history that a character more akin to the IT Crowd than James Bond was capable—undetected and unscathed—of torpedoing the most ambitious spy program ever conceived.)

But the explanation is no less baffling than the plot. Fed up with what he considered injustice, he decided to take it to the people to spark a national dialogue about privacy. That might have been OK if we hadn't already played this game before. We played it in 2005 when it was leaked that Bush had circumvented FISA [Foreign Intelligence Surveillance Act], breaking federal law countless times by warrantlessly wiretapping Americans. After [the Bush administration] reassuring the nation that such surveillance only hit foreign correspondence and was neither recorded nor used for any purpose other than combatting terrorism, in 2006 and 2008, we learned that both claims were false. Snowden's justification is hollow, as America already knew about these programs, and its response has repeatedly been a collective "Meh."

But this is all beside the point. Any defense of Snowden ended when he kept talking. And, as if hell-bent on crushing all hope of redemption, he keeps providing details.

Only a cartoonishly childlike understanding of diplomacy could conclude that detailing our spying on our allies doesn't hurt our national security. To defend Snowden on the grounds he has avoided disclosing information concerning terrorism, as some have, is as myopic a view of national security as thinking that avoiding U.S. casualties by using drones to kill terrorists already willing to die for their beliefs—adjacent innocents be damned—is the best way to combat terrorism. Snowden handed the [British newspaper] *Guardian* and [the German newspaper] *Der Spiegel* exquisite details. How does it further a conversation regarding privacy to reveal that the U.S. intercepted 85,489 texts from [Mexico's president] Peña Nieto's office, or that it illegally tracked financial transactions and bugged European Union offices? Now what happens if the U.S. wants to renegotiate trade with Mexico, or needs EU nations' help making environmental policy in the UN? Snowden just made everything incredibly more difficult.

Of course allies spy on each other, just as couples fight. Adults try not to fight publicly so that people don't get the wrong idea, let alone broadcast it to billions. Whether the outrage of world leaders is genuine or Renault-esque, in the face of such public embarrassment they must save face. This usually means some form of action against our government. Explain again how this helps the interests of the United States, or that precious dialogue which Snowden hopes to facilitate.

The continued disclosure of details belies Snowden's oft-repeated claim that this is all for the greater good. If we already knew that we were data-mining inconceivable amounts of information, not just on Americans but on everyone everywhere, the revelation of exactly on whom and by what means we spied offers delicious tidbits without providing any useful insights. Hearing how many of Felipe Calderon's [president of Mexico, 2006–2012] emails the NSA [U.S. National Security Agency] hacked is not news, but porn—and at this point anyone who still compares Snowden to Franklin might as well assert that TMZ [a celebrity news/gossip website] wields the same gravitas as *The Economist* or *Foreign Policy*. The important question has never been whether Snowden is a selfless patriot or a narcissistic nihilist. It is how we convince him to stop.

WILLIAM H. HARWOOD teaches philosophy at the University of Texas–Pan American, specializing in political philosophy and ancient philosophy. He is also a board member of Planned Parenthood Association of Hidalgo County and a Battleground Texas Volunteer Organizer.

Emily Bazelon  **NO**

Is Edward Snowden a Traitor? If He Is, So Was Daniel Ellsberg

The condemnations are raining down upon Edward Snowden, master leaker of National Security Agency surveillance programs. They come from the expected sources: House Speaker John Boehner calls him a "traitor." But they also come from people you might expect to be more sympathetic toward him. Legal experts Geoffrey Stone of the University of Chicago and Jeffrey Toobin of the *New Yorker* both believe he betrayed his country and should go to prison. So does *New York Times* columnist David Brooks.

Toobin writes that Snowden "wasn't blowing the whistle on anything illegal; he was exposing something that failed to meet his own standards of propriety." Stone says, "There is no reason on earth why an individual government employee should have the authority, on his own say so, to override the judgment of the elected representatives of the American people and to decide for the nation that classified information should be disclosed to friends and enemies alike."

Brooks: "He betrayed the Constitution. The founders did not create the United States so that some solitary 29-year-old could make unilateral decisions about what should be exposed." (Brooks also dings Snowden for being a lone wolf, which is hilariously at odds with his adulation of other solo radicals.)

A foundation of their argument is that Snowden is not a genuine whistle-blower. And it's true that if you divulge classified information to expose the government and you don't reveal a clear legal violation, you're not, under current law, a whistle-blower. The federal Whistle-blower Protection Act, passed in 1989, was written to shield government employees who reveal fraud and other wrongdoing. But it is riddled with exceptions. If you work for the NSA or the CIA, you're out of luck—no protection for you. Snowden misses on both counts: He seems to have exposed no actual crimes, and he worked for the NSA. They may make movies about private-sector whistle-blowers such Erin Brockovich (fought toxic dumping) and

Jeffrey Wigand (exposed tobacco company lies). But they prosecute government whistle-blowers such as Thomas Drake, who exposed waste and bureaucratic mess at the NSA, and of course Bradley Manning, on trial for the enormous WikiLeaks data dump.

The Obama administration is moving to charge Snowden with disclosing classified information, probably under the Espionage Act—the anti-sedition law from 1917 that has recently become the government's favorite weapon. The government can count on this much: Once Snowden is charged with crimes that will surely carry a long prison sentence, it will be harder to see him as a hero. That has certainly been true for Manning.

I understand that some government secrets must stay secret. And I recognize that the decision of one employee to reveal what many of his superiors have ruled classified is inherently troublesome for people in power. And I also know that whistle-blowers are rarely pure and innocent and lovable. They are often paranoid and obsessive. They fill up your inbox with confusing and demanding messages. They ask you to hack through thickets of confusing material. In other words, they are time-consuming zealots who are often trying to confront you with knowledge you wish wasn't true.

But here's my question for Snowden's denouncers: What about Daniel Ellsberg? It's only in retrospect that the leaker of the Pentagon Papers has acquired the status of a national icon. [The Pentagon Papers are the informal name of a internal U.S. government study of U.S policy toward Vietnam between 1945 and 1967. Ellsberg was an analyst with the Rand Corporation, a private research organization working on the study. He photocopied confidential government documents in Rand's possession and gave them to the *New York Times* and other media outlets.] At the time he leaked the documents in 1969, Ellsberg was a man with top security clearance who was accused of betraying his government by exposing its greatest secrets. About an actual, troops-on-the-ground war. The Pentagon

Papers showed that the [President Lyndon B.] Johnson administration knew that casualty figures in Vietnam would be much higher than the numbers it had publicly projected. Also, that the Johnson administration had lied about the war to Congress.

Looking back 40 years, we treat harsh criticism of the Vietnam War as patriotic and foresighted. But Ellsberg must have sounded like any draft-dodger when he decided, as he wrote in his memoir, that the war was "mass murder." And he took it upon himself not to protest through any proper channels, but to go through the press. Ellsberg first took the Pentagon Papers to some senators and asked them to put it into the congressional record. He went to the press after they refused. When Arthur Ochs Sulzberger, the publisher of the *New York Times* at the time, died last year, his obituary in his own newspaper called publishing the Pentagon Papers excerpts "a defining moment for him and, in the view of many journalists and historians, his finest." But at the time, as [columnist] Anthony Lewis pointed out, "it was not an easy decision for the *Times*" to publish.

No wonder, then, that Ellsberg too was prosecuted for breaking the law under the Espionage Act. "I expected to go to prison for life," he told NPR [National Public Radio] in 2011. The charges against him were dropped only because the [President Richard M.] Nixon administration egregiously violated his rights by breaking into his psychiatrist's office to look for information that would embarrass him. "The bizarre events have incurably infected the prosecution of this case," the judge wrote.

If Nixon's men hadn't gone so completely off the reservation, how would we think about Ellsberg today? And how does his whistle-blowing stand up to the test posed by Toobin, Stone, and Brooks? Ellsberg and Snowden both acted unilaterally. They both ignored the collective judgment of the country's duly elected representatives. I suppose you could argue that Ellsberg was exposing illegalities, if the Pentagon Papers showed Johnson administration officials lying to Congress. But if that's the main test for being a whistle-blower, Snowden's revelations seem to catch Director of National Intelligence James Clapper in misleading testimony, too. I realize we can't have rogue operators undermining the government at every turn. But I keep coming back to this question: How do we make room for the secret-tellers who only history can show were on the right side?

EMILY BAZELON serves as senior editor of *Slate* magazine and is coeditor of DoubleX. Ms. Bazelon edits "Assessment" and the magazine's health and science sections. Before joining *Slate*, Ms. Bazelon worked as an editor and writer at *Legal Affairs* magazine and as a law clerk in Portland, Maine. Her work has appeared in the *New York Times Magazine, Mother Jones, The Washington Post*, the *Los Angeles Times*, and the *Boston Globe*. Ms. Bazelon is a graduate of Yale College and Yale Law School.

EXPLORING THE ISSUE

Has Edward Snowden Damaged the U.S. Public Interest?

Critical Thinking and Reflection

1. Which is more important: guarding national security secrets or ensuring the right of Americans to know what their government is doing so that citizens can exercise their democratic authority to control the government?
2. There is little doubt that Edward Snowden violated U.S. law, at least technically. If it is also true that some of his disclosures revealed illegal practices by the National Security Agency, does that fact mitigate or even obviate his illegal actions? Bear in mind that only a small fraction of the documentation Snowden disclosed had anything to do with the NSA.
3. Many newspapers and other news media, as well as numerous Americans, believe that the president should grant Edward Snowden a pardon and allow him to return to the United States without fear of prosecution. Do you agree?
4. Should Edward Snowden's motives make any difference in judging his actions?
5. If Snowden violated the law by copying and distributing classified documents (as was arguably true for Daniel Ellsberg), then do you agree with the Supreme Court in its decision in *New York Times v. U.S.* (1971) that it is legal for the news media to print and broadcast information that is both classified and illegally obtained?

Is There Common Ground?

The issues surrounding Edward Snowden, Bradley Manning, Daniel Ellsberg, and others in U.S. history center on several conundrums. The first relates to the right of the people to know versus the legitimate necessity of the government keeping some secrets related to national security, international diplomacy, and other matters. The answer would be simple if the government kept few secrets and only did so when revealing them presented a "clear and present danger" to the safety and well-being of the country. The reality, though, is that the government classified a huge amount of material (over 95 million documents in 2012 alone). Moreover, past leaks have shown that at least part of these documents are classified to keep U.S. illegal or dubious practices, blunders, and other politically embarrassing material from becoming public.

A second conundrum relates to the fact the disclosures, especially of the massive variety such as Snowden's, often contain information the public should rightfully know (the extent to National Security Administration communications intercepts) and also information that damages the national interest and perhaps even puts U.S. covert operations and personnel in danger. Snowden and his supporters have focused on the NSA material, but as

William Harwood points out in the first article, that material was only a small part of what was made public.

The third conundrum relates to what to do with the Ellsbergs, Mannings, and Snowdens of life. Arguably their disclosures did some public good, but it is a stretch to argue that any one of them acted out of pure patriotism and regard for democracy. Each disagreed with policy, and, less certainly, each was also driven by one or more motives such as self-promotion and other emotional ego factors. What is the line between noble citizen and traitor?

Create Central

www.mhhe.com/createcentral

Additional Resources

Ellsberg, Daniel. (2003). *Secrets: A Memoir of Vietnam and the Pentagon Papers* (Penguin Books)

A memoir by the individual who made the first "great leak."

Greenwald, Glenn. (2014). *No Place to Hide: Edward Snowden, the NSA, and the U.S. Surveillance State* (Metropolitan Books)

This book by an investigative reporter for the British publication *The Guardian*, who has been sympathetic to Snowden, claims to contain new information on the extent and impacts of government surveillance programs, both domestically and abroad.

Grier, Peter. (December 21, 2010). "How Many Documents are Classified Exactly? That's a Secret. But Here's an Educated Guess." *Christian Science Monitor* article located at http://www.csmonitor.com/USA/DC-Decoder/Decoder-Wire/2010/1221

This article takes up the degree to which the U.S. government classifies material. The government reports "classification decisions," but this standard can include groups of related documents, each of which has many pages.

The Washington Post (2013). *NSA Secrets: Government Spying in the Internet Age* (Diversion Press)

A brief e-book that is a compilation of reporting by *The Washington Post* about the National Security Agency surveillance programs. It is available for $2.99 on a variety of e-book sources, including Amazon.

Internet References . . .

The National Security Agency

One element of the U.S. intelligence structure and the focus of much of the controversy over the disclosures contained in the documents leaked by Edward Snowden.

http://www.nsa.gov/

The NSA Files

Information on the website of *The Guardian*, a decidedly liberal, but not left-wing British newspaper. It is published in Manchester, UK, but has a significant international following. The site contains information on the NSA, with an emphasis on the information disclosed by Edward Snowden's leaks and the unfolding of the story.

http://www.theguardian.com/world/the-nsa-files

The Washington Post

The newspaper maintains a site that follows the Snowden affair and is thus good for updates and background information. Unfortunately, like most sources, *The Washington Post* focuses on disclosures about the NSA and does little regarding the information in the much larger number of documents related to other aspects of U.S. foreign policy and national security.

http://www.washingtonpost.com/nsa-secrets/

WikiLeaks

WikiLeaks is a primarily cyberspace organization legally anchored in Sweden that describes itself as "a not-for-profit media organization" whose "goal is to bring important news and information to the public" by providing "an innovative, secure and anonymous way for sources to leak information to our journalists." WikiLeaks says that "one of our most important activities is to publish original source material alongside our news stories so readers and historians alike can see evidence of the truth."

http://wikileaks.org/